Catalogue of Plans of Proposed Canals,
Turnpike Roads, Railways and other Public Works
deposited with the Clerks of the Peace for
Northamptonshire, the County Borough of Northampton
and the Soke of Peterborough, 1792–1960,
in the Northamptonshire Record Office

Compiled by Philip Riden

*County Editor
Victoria County History
of Northamptonshire*

Northamptonshire Record Office

First published 2000

Produced for the Northamptonshire Record Office
Wootton Hall Park, Northampton NN4 8BQ
by Merton Priory Press Ltd
67 Merthyr Road, Whitchurch
Cardiff CF14 1DD

ISBN 1 898937 43 5

© Northamptonshire Record Office 2000

Printed by Technical Print Services Ltd
Brentcliffe Avenue, Carlton Road
Nottingham NG3 7AG

CONTENTS

Preface   4

Introduction   5

Northamptonshire Canal Plans   19

Northamptonshire Turnpike Plans   23

Northamptonshire Main Series Plans: Summary   25

Northamptonshire Main Series Plans: Catalogue   35

Soke of Peterborough Plans   78

Index   81

PREFACE

When work re-started on the Northamptonshire Victoria County History in 1996 it was agreed that a general volume dealing with transport, communications and modern industry should be one of the first to be prepared. When I began to collect information for the sections dealing with canals, railways and turnpike roads, it occurred to me that a useful preliminary to writing those chapters would be the compilation of a more detailed catalogue than was then available describing the plans (and other documents) deposited with the clerks of the peace for Northamptonshire, the Soke of Peterborough and the County Borough of Northampton which are now in the Northamptonshire Record Office, and that such a catalogue might merit publication in its own right.

This booklet is the outcome of the work which I have been doing on the Northamptonshire deposited plans over the last two years. I am very grateful to the County Archivist, Miss Rachel Watson, for agreeing that I should undertake this project, and to her staff for producing in the searchroom large numbers of often rather bulky plans. I am also indebted to the Northamptonshire Victoria County History Trust for their support for this by-product of my work for them and in particular to Mr Roger Morris, the Chief Executive and Town Clerk of the Borough of Northampton, for reading and commenting on an earlier draft of the introduction. Miss Watson has also read the work in draft.

In tackling this project I have obviously followed to some extent the models provided by similar catalogues (both published and unpublished) of deposited plans in other record offices and I am grateful to a number of county archivists for kindly supplying me with examples of their lists. Partly because Northamptonshire has a relatively modest number of plans, compared with more heavily industrialised counties, I decided that it was possible, as well as desirable, to describe each plan in some detail, including the names of all the parishes through which a particular scheme was to run, not merely those in Northamptonshire. This approach, combined with a fairly full introduction, may give the work some wider value outside Northamptonshire.

University College Philip Riden
Northampton August 2000

INTRODUCTION

The class of county (or borough) records familiarly and conveniently known as 'Deposited Plans' (the term used here) or less accurately as 'Railway Plans' in fact includes several other documents as well as the plans themselves and relates to a wider range of public works than railways alone. They are also of greater value than is sometimes appreciated for the investigation of topics other than transport history.

The evolution of the deposited plan

Although legislation by private (as opposed to public) Act of Parliament has much older origins, the creation of a detailed code of procedure for private bills, including the deposit of plans and other documents, began only in the second half of the eighteenth century.[1] In 1774 the House of Commons appointed a committee whose report formed the basis of seven standing orders requiring the submission of certain information by the promoters of bills for inclosing, draining or improving lands, fens or commons, and for making turnpike roads, canals or river navigations, to which the House added an eighth, requiring the promoters' compliance with the other seven to be examined by a committee, thus establishing the principle that private bills should be examined in this way.[2] In 1786 bills for building bridges were brought within the scope of the standing orders of 1774.[3] Three years later, as the volume of canal bills increased, new standing orders were made concerning the notice to be given by promoters to quarter sessions. In 1792, at the beginning of the main period of the Canal Mania, more detailed standing orders were recommended for such bills, requiring the publication of notices in the *London Gazette* and local newspapers, such notices to name the parishes through which the canal would pass, and the deposit (on or before 30 November in the year preceding the intended application to Parliament) of a map or plan and book of reference with the clerk of the peace or town clerk of a quarter sessions borough in England (the principal sheriff clerk in Scotland), who was to make a memorial on the plan and book of reference noting the date and time at which they were deposited. Clerks were to permit inspection of the deposited plan and book of reference, on payment of a fee, at all reasonable times. Compliance with these standing orders was to be enquired into by the committee considering the bill. These recommendations were made standing orders on 7 June 1792 and

---

[1] The two principal general studies of the subject remain F. Clifford, *A History of Private Bill Legislation* (1885–7) and O.C. Williams, *The Historical Development of Private Bill Procedure and Standing Orders in the House of Commons* (1949), of which the latter forms the basis of the following account; for procedure in the House of Lords see M.F. Bond, *Guide to the Records of Parliament* (1971), 70–82, 85, 89–90. See also J. Simmons and G. Biddle, *The Oxford Companion to British Railway History* (1997), 368–9, 400–1.

[2] Williams, *Historical Development*, i. 264.

[3] Ibid., i. 265.

ordered to be printed, with copies sent to clerks and sheriffs.[1]

These standing orders are the origin of the deposit of plans of public works of all kinds with local officials. They were revised in 1794, when an order was made requiring the deposit of a duplicate plan and book of reference with the House of Commons,[2] extended in 1799 to bills relating to railways or dramroads (i.e. tramroads),[3] and again in 1807 to include turnpike road bills.[4] In 1810 the House of Commons appointed a committee to consider all aspects of private bill procedure, which led to the establishment of a Private Bill Office and the publication, for the first time, of all the private bill standing orders.[5] Another committee in 1810–11 prepared 19 new or revised standing orders, including some relating to the scale of the deposited plans, the deposit of sections as well as plans, and the deposit of duplicates of both, as well as other documents, with local officials and the Private Bill Office. As revised, the standing orders were again printed in full and copies sent to clerks and sheriffs.[6] In 1812–13 new standing orders relating to canal and navigation bills prescribed a scale for plans of not more than 5 and not less than 3 inches to the mile, in place of the previous requirement for plans at a scale of 1 inch to the mile. The orders for such bills were revised again the following session.[7]

In 1830 the standing orders were reprinted afresh but soon required extensive revision as large-scale railway construction got underway. A Select Committee was appointed to consider the question in 1836, which reported in February the following year, when new standing orders were made. Another Select Committee was appointed the same day as the first reported, specifically to consider standing orders for railway bills, which reported in August 1837, when further new standing orders were made. The main changes affected the date of deposit of plans for railways (but not other schemes), which was moved back from 30 November to 1 March in the year preceding application to Parliament, and the scale at which the plans were to be drawn.[8] The change in date proved inconvenient and was abandoned in 1842 in favour of 30 November.[9]

During the Railway Mania of 1844–6 and in its aftermath numerous further changes were made to the standing orders, almost all arising from railway promotion, and it was during this period that most of the detailed orders concerning the preparation of the deposited plan were made.[10] The Railway Clauses Consolidation Act of 1845 also made some new rules concerning the plan and book of reference.[11] From 1837 the scale of the plan was to be not less than 4 inches to the mile and it was to show all the land that was to be taken and also the limit of deviation within which the railway might be built either side of the proposed route. Unless the whole plan was drawn on a scale of not less than a quarter of an inch to 100 feet (1:480), enlarged plans were to be drawn on such a scale of any buildings within the limit of deviation.[12] The plan was to show the distance in miles and furlongs from each

---

[1] Ibid.

[2] Ibid., i. 44 (but cf. i. 266).

[3] Ibid., i. 266.

[4] Ibid., i. 268.

[5] Ibid., i. 268–9.

[6] Ibid., i. 269–70.

[7] Ibid., i. 270.

[8] Ibid., i. 60–4.

[9] Ibid., ii. 39–42; G. Biddle, *The Railway Surveyors. The Story of Railway Property Management 1800–1990* (1990), 41–3.

[10] Williams, *Historical Development*, i. 66–7.

[11] 8 Vict. c. 20, ss. 7–15; Biddle, *Railway Surveyors*, 44–5.

[12] Williams, *Historical Development*, ii. 67–8.

terminus, and the radius of every curve under one mile in length. Tunnels were to be shown in dotted lines and no work was to be shown as tunnelling where it would be necessary to cut through or remove surface soil[1] (i.e. cut and cover tunnelling). Any highway diversions, including footpaths, were to be shown on the plan.[2] From 1846 the large-scale deposited plan had to be accompanied by a one-inch Ordnance Survey map on which was marked the general route of the railway or whatever.[3]

The section was to be drawn to the same horizontal scale as the plan and to a vertical scale of not less than 1 inch to 100 feet (1:1200). It was to show the surface of the ground marked on the plan, the intended level of the proposed work, the height of every embankment and the depth of every cutting. It was to have a horizontal datum point which had to remain constant and be related to a fixed point near some portion of the work (in the case of a canal or railway, near either terminus). The distance of the fixed point above or below an Ordnance Survey benchmark in the neighbourhood was to be stated.[4] In the case of railways, the line shown on the section was to correspond with the upper surface of the rails.[5] Distances along the horizontal datum were to be marked in miles and furlongs corresponding with those on the plan, and a vertical measure from the datum line to the railway was to be marked at the beginning and end of the railway and at every change in gradient; between each change the inclination of the gradient was to be shown.[6] Where a railway was to cross any public road, river etc., its height or depth above or below was to be shown, together with the height and span of every arch. Any change of existing levels at level crossings was to be indicated.[7] Where an embankment was pierced by a bridge or viaduct of more than three arches, or a cutting had a section of tunnel within it, the extreme height or depth of the embankment or cutting was to be marked on each part that was divided by other works.[8] Any change in the water level of a canal, or the gradient of a road or railway, was to be stated on the section and drawn in cross-section at a horizontal scale of not less than 1 inch to 330 feet (1:3960) and a vertical scale or not less than 1 inch to 40 feet (1:480). All such cross-sections were to extend 200 yards either side of the centre-line of the railway.[9]

The book of reference was to contain the names of the owners, lessees and occupiers of all land and houses which might be taken, together with a full description of the premises.[10]

The general principles of local deposit did not alter during the 1830s and 1840s, apart from the short-lived change of date of deposit between 1837 and 1842. From the former year a duplicate plan and section had to be deposited with the clerk of the peace, which was to be sealed up and kept until sent for by either House of Parliament; the other copy remained open to inspection.[11] In 1847 the days on which deposits might be made were restricted by excluding Sunday, Christmas Day, Good Friday and Easter Monday, and the hours on other days limited to 8 a.m. to 8 p.m.[12] In 1837 the rule which had applied to canal and river

---

[1] Ibid., ii. 70–1.
[2] Ibid., ii. 71.
[3] Ibid., ii. 39–42.
[4] Ibid., ii. 74–5.
[5] Ibid., ii. 75.
[6] Ibid., ii. 76.
[7] Ibid., ii. 76–7.
[8] Ibid., ii. 78.
[9] Ibid., ii. 77–8.
[10] Ibid., ii. 73.
[11] Ibid., ii. 39–42.
[12] Ibid., ii. 39.

plans since 1814, requiring the deposit with parish officials of extracts from the plan, section and book of reference relating to land and buildings in the parish in question, was extended to all types of plan, with a date of deposit of 1 April for railways and 31 December for other schemes. In 1846 the date was changed to 30 November, the same as the date for deposit with the clerk of the peace.[1]

By 1851, when the standing orders were again reprinted, a complete code had finally been achieved, much of it since 1835 under pressure from railway development.[2] Numerous textbooks for surveyors, parliamentary agents and railway promoters were also published in this period.[3] Most later changes resulted from the widening scope of private bill legislation and the growth of the provisional order system, to include (from 1870) street tramways (and from 1889 'tramroads' in the modern parliamentary sense, meaning passenger tramways running on a reserved track, rather than the highway),[4] as well as water and gas undertakings from 1847,[5] and the generation and supply of electricity from 1882.[6] Plans also had to be deposited for schemes promoted under the Light Railways Act of 1896,[7] although not for lines built for purely private use, whether to carry goods and minerals (such as the extensive network of standard and narrow gauge lines which once served the Northamptonshire ironstone industry)[8] or passengers (such as the miniature railways in public parks or on landed estates built by the Northampton engineer W.J. Bassett-Lowke).[9]

The only later nineteenth-century change relating to the detailed preparation of plans and sections came in 1865, at the suggestion of an engineer, T.E. Harrison, when a new standing order required that where a proposed railway was to make a junction with an existing or authorised line, the course of the latter was to be shown on the plan for 800 yards on either side of the proposed junction on the same scale as the rest of the plan, and its gradients over the same distance on the section.[10]

The standing orders were revised in 1929–30, when the date for local deposit was changed from 30 November to 20 November.[11] More radical reform was undertaken in 1942 and 1945, although as the head of the Committee and Private Bill Office at the time observed, some archaic features survived even after this date, including the requirement for the clerk to 'make a memorial' on the documents deposited with him.[12] Since, under the 1888 Local Government Act, the documents were from 1889 deposited (in England) with the clerk of the county council or town clerk of a county borough, the abolition of the office of clerk of the peace under the Courts Act of 1971 did not affect the system, although by that date very few deposits were being made. Only with the passing of the Transport and Works Act of 1992, which was intended to create new arrangements for the promotion of railways (including tramways, light railways etc.), inland waterways and works interfering with rights of

---

[1] Ibid., ii. 49–50.

[2] Ibid., i. 117–18.

[3] Biddle, *Railway Surveyors*, 46–8, 65.

[4] Clifford, *History*, 188; Williams, *Historical Development*, ii. 72–3.

[5] Clifford, *History*, 221–3.

[6] Ibid., 236.

[7] *Oxford Companion to British Railway History*, 263–5.

[8] For which see E. Tonks, *The Ironstone Quarries of the Midlands* (Cheltenham, 1988–92).

[9] For which see R. Fuller, *The Bassett-Lowke Story* (1984); J. Bassett-Lowke, *Wenman Joseph Bassett-Lowke. A Memoir of his Life and Achievements* (Chester, n.d. [1999]).

[10] Williams, *Historical Development*, ii. 71, 79.

[11] Ibid., ii. 29–30.

[12] Ibid., i. 46.

navigation, closer to that which had long applied to major road projects,[1] did the practice of depositing plans with local authorities on the lines introduced two hundred years earlier come to an end, just at a time when schemes for large-scale railway construction in Great Britain were being seriously discussed for the first time since before 1914 and there was renewed interest in light rapid transit systems.

Traditionally, deposited plans of any date have been treated as part of the records of the court of quarter sessions for the county in question. Quarter sessions records are public records within the meaning of the Public Records Act of 1958, for which a place of deposit other than the Public Record Office has been specified, i.e. a local record office appointed under the Act. When county councils (and county borough councils) were established under the Local Government Act of 1888 the clerk of the peace of each county was also made clerk of the county council,[2] although the two offices remained legally distinct. However, the Act specified that the clerk of the peace, when acting 'under the Acts relating to ... the deposit of plans or documents ... shall act under the direction of the county council, and all enactments relating to such ... deposit, shall be construed as if clerk of the county council were therein substituted for clerk of the peace'.[3] This sub-section seems to mean that plans deposited on or after 1 April 1889, the day on which county councils came into existence, were deposited with the clerk of the county council, not the clerk of the peace.[4] If this is the case, such plans are not public records, but local government records. Those deposited with town clerks in county boroughs from that date certainly appear to be local government records. In practice, most local record offices make no distinction between material deposited before or after that date, and in Northamptonshire the clerk continued to sign memorials on plans for railway and canal works (although not gas, water or electricity schemes) as clerk of the peace, rather than clerk of the county council, until at least 1960 (see **300–302, 316–18**). Under the 1972 Local Government Act plans were deposited with the 'proper officer' of the local authority.[5]

The Records

By the late 1840s, and in most respects by the beginning of the 1830s, the documents deposited with the clerk of the peace (or town clerk) had settled into a standard form, which subsequently changed little until the system came to an end. A complete deposit would normally include (in duplicate):

(a) a series of lithographed plans showing the line of the intended railway (or whatever) and a strip of land either side within which the route, as finally built, might deviate, together with inset plans on a larger scale of built-up areas;

(b) a corresponding series of longitudinal sections and cross-sections;

(c) an index map, made up from Ordnance Survey 1:63,360 sheets where these were available, on which the route of the line was highlighted in coloured ink;

---

[1] *Oxford Companion to British Railway History*, 401.

[2] 51 & 52 Vict. c. 41, s. 83(1).

[3] 51 & 52 Vict. c. 41, s. 83(6).

[4] Williams, *Historical Development*, ii. 39–43 discusses the development of the standing order governing local deposit, which as revised in 1942 referred to the clerk of the county council, not the clerk of the peace, without giving a date at which the change was made. The Act of 1888 appears to effect the change.

[5] 20 & 21 Eliz. II c. 70, s. 225.

(d) a book of reference, normally letterpress printed, scheduling every parcel of land which lay within the limits of deviation shown on the plans, arranged by parish, listing the owner, lessee and occupier, with a description of the land; and

(e) a print of the notice published (for schemes in England) in the *London Gazette* setting out the main provisions of the proposed bill, including a written description of the line, listing all the parishes through which it was to pass.

The index map, plans and sections were usually bound in a card cover, on which was printed the name of the scheme and its engineer (and sometimes also the surveyor), as well as the lithographer. The set of plans seems normally to have been submitted to the clerk of the peace rolled up, with the book of reference loosely inserted inside and the *Gazette* notice pasted or stitched inside the front cover.

In almost all cases, promoters deposited with every local authority through whose area a scheme passed a complete set of documents covering the entire route of the railway, not merely the section within the county or borough in question, although sometimes plans relating to branch lines which were part of a larger scheme but were themselves situated entirely in one county might only be deposited with the clerk of that county.

If the project was successful, the Act would receive the Royal Assent in the summer of the year following that in which the plans were deposited.[1] The printed plan covers are sometimes dated with the year in which they were deposited (and possibly also the month); other promoters used the date of the parliamentary session in which the bill was to be introduced, which for the earlier plans would generally be the following year, since until the second half of the nineteenth century Parliament did not always sit before Christmas. In this catalogue, the year in which the plans were deposited has been used throughout.

When they were deposited, the plans were endorsed with what the standing orders described as a 'memorial', recording not only the date of deposit but also the hour of the day, so as to indicate that the plans had been deposited before the deadline of midnight on 30 November. At the height of the Railway Mania in 1845, vast numbers of plans were deposited right up to the deadline (and beyond) at clerks' offices all over England, which had to open for business on 30 November that year, even though it was a Sunday.[2] Some 36 plans were deposited with the Northamptonshire clerk of the peace that day (about 15 per cent of all the railway plans deposited over more than a century), some of which are noted as having arrived shortly after midnight. If opponents of the bill discovered this, they could secure its rejection on the ground that the promoters had failed to comply with standing orders. There is only one plan in the Northamptonshire series (**9**) deposited during the period in which the deadline was 1 March.

The plans deposited with the two Houses of Parliament are available for consultation in the House of Lords Record Office. The Commons plans (and books of reference) prior to 1819 are lost and in 1904 many plans which were identical to those in the Lords series were destroyed (although in some cases a Commons plan may fill a gap in the Lords series). The House of Lords plans and books of reference begin in 1794. Both series continue to the present day, since the deposits are made under standing orders of the two Houses and the procedure was therefore not affected by the Transport and Works Act of 1992.[3]

---

[1] Biddle, *Railway Surveyors*, 39–41.

[2] Ibid., 51.

[3] Bond, *Records of Parliament*, 228–9 (Commons), 85, 89–90 (Lords), and additional information supplied by the House of Lords Record Office, which advises me that the statement in *Oxford Companion to British Railway*

In 1845 a new standing order also required the deposit of plans for railways with the Railway Department of the Board of Trade (or, for a short time after 1846, the Railway Commissioners).[1] Similarly, from 1884 tramway plans had also to be deposited with the department and from 1900 maps showing the proposed supply area of electricity schemes.[2] The Railway Department became part of the Ministry of Transport in 1920 but the deposited plans have not survived to be transferred to the Public Record Office.[3] Plans for particular types of scheme also had to be deposited with other departments, including the Home Office (where churchyards were to be taken), the Admiralty (bills affecting tidal lands), the Board of Agriculture and fisheries (bills affecting fisheries or common land), and the fisheries and Harbours Department of the Board of Trade (bills affecting the banks of a river),[4] but in no case have the plans been retained.

Very few (if indeed any) of the extracts from plans and sections relating to a particular parish, which had to be deposited with parish clerks under a standing order of 1814 relating to canals and rivers and another of 1837 applying to all types of scheme (which was later extended to non-county boroughs and urban and rural district councils),[5] appear to have survived to reach county record offices.[6]

Deposited plans in the Northamptonshire Record Office

The Northamptonshire Record Office (NRO) is appointed as the place of deposit for the records of three commissions of the peace, Northamptonshire, the Soke of Peterborough and the Borough of Northampton. Under the Local Government Act of 1888 separate county councils were established for Northamptonshire and the Soke,[7] of which the former continued until 1974; the administrative county of the Soke of Peterborough was combined with Huntingdonshire in 1965 to form the county of Huntingdon & Peterborough, which was itself merged with Cambridgeshire into a new administrative county (known as Cambridgeshire) in 1974.[8] Northampton was a county borough between 1889 and 1974.[9]

Deposited plans from Northamptonshire and Peterborough quarter sessions records, from Northamptonshire County Council and from Northampton County Borough Council have been transferred to the NRO. Those from Northamptonshire form by far the largest group and include plans for a number of schemes which lie entirely within the Soke or the Borough. The smaller (and less well preserved) series transferred from the Soke contains only one plan that

---

History, 369 that the HLRO holds only plans for schemes that were enacted is incorrect: the office retains all plans deposited with the two Houses. See also M.F. Bond, 'Materials for Transport History amongst the Records of Parliament', *Journal of Transport History*, iv (1959–60), 37–52; H.S. Cobb, 'Parliamentary Records relating to Internal Navigation', *Archives*, iv (1969), 73–9.

[1] Williams, *Historical Development*, ii. 47–8; for the development of the department see H. Parris, *Government and the Railways in Nineteenth-century Britain* (1965).

[2] Williams, *Historical Development*, ii. 44.

[3] *Guide to the Contents of the Public Record Office* (1963), ii. 277, 280; D.B. Wardle, 'Sources for the History of Railways at the Public Record Office', *Journal of Transport History*, ii (1955–6), 214–34.

[4] Williams, *Historical Development*, ii. 45–7.

[5] Ibid., ii. 49–50.

[6] This comment is based on my own experience of parish collections in a limited number of record offices and on speaking to an unsystematic sample of archivists elsewhere. It may not be true of all counties.

[7] 51 & 52 Vict. c. 41, s. 46(1)(d).

[8] F.A. Youngs, *Guide to the Local Administrative Units of England*, i (1979), 628–30.

[9] 51 & 52 Vict. c. 41, sch. 3; 20 & 21 Eliz. II c. 70.

was not also deposited with the clerk of the peace for Northamptonshire. The Northampton deposited plans have not been consulted in the preparation of this catalogue, but by analogy with those for the Soke it seems unlikely that the Borough series includes any which are not also available among the county records. Most of what follows, therefore, relates to the plans deposited with the clerk of the peace for Northamptonshire.

The Northamptonshire deposited plans

The earliest plans in the county series endorsed with a date of deposit are those for the Grand Junction Canal between Braunston and Brentford, deposited on 10 November 1792, and for the Leicestershire & Northamptonshire Union Canal between Leicester and Northampton, deposited the following day; there are two other plans for the latter scheme dated 1792 but without a memorial, an undated plan of the Grand Junction Canal which presumably dates from 1792, and a plan dated 1792 for the Ashby de la Zouch Canal, no part of which lay in Northamptonshire. From here the series transferred to the county record office continues until 18 November 1960, when a water scheme for the River Great Ouse was deposited. As well as 221 railway plans, the series includes 17 plans for canals, a similar number for turnpike roads, 11 relating to river drainage or navigation, 24 plans from 1859 onwards for gas undertakings, 24 plans for water schemes from 1864, 22 for electricity schemes from 1889, 13 for tramways or tramroads, and five relating to general town improvement. In 1931 a plan was deposited under the Petroleum (Consolidation) Act 1928 (**303**) showing the location of filling stations in the county and, incidentally, scheduled ancient monuments and war memorials, since petrol could not be stored within a certain radius of either. The last canal plans were deposited in 1930–1 (**300**, **302**) showing minor improvements to the former Grand Junction Canal that followed the creation of the Grand Union; the last railway plan (for a footpath diversion at Blisworth) was deposited by the British Transport Commission in 1954 (**317**); and the last gas schemes date from 1942–3 (**314**, **315**).

The present arrangement of the Northamptonshire plans appears to have been created in 1902, when the clerk divided them into three series, one containing canal and river plans dating from 1792 to 1836, a second the turnpike road plans (1809–39), and a third all other plans, starting with the earliest proposal for a railway from London to Birmingham, which was deposited on 30 November 1830. This division was presumably based on the (archivally mistaken) view that the early canal and turnpike plans formed separate (closed) classes, whereas what may have been thought of as the 'railway plans' were still accruing (although this group also contains river drainage and navigation plans dating from after 1836 and would, after 1902, have two further canal plans added to it). Each of the three groups was given a separate set of serial numbers: 1–25 for the canals, 1A–15A (together with 8B and 10B) for the turnpike roads, and 1–250 for the railways and other schemes included in the main series. None was arranged in strictly chronological order. Each separately numbered item was wrapped in brown paper, to which a label bearing a handwritten serial number and a brief description of the plan was pasted. From No. 251 in the main series (deposited on 30 November 1902) the numbering was applied using a stamp, which makes it possible to date the creation of the present arrangement with some confidence. None of the plans bears any evidence of an earlier numbering system.

For the period during which a second, sealed, set of documents had to be deposited, the same method of wrapping and numbering as that used for the set open for inspection seems to have been adopted, both before and after 1902, although occasionally after this date both sets will be found in a single wrapper. Generally, the treatment of twentieth-century deposits was less formal than in the nineteenth century, especially for schemes other than railways and canals. An ordinary office date-stamp was used to record the date of deposit, rather than a handwritten memorial, and sometimes correspondence between the clerk and the promoters or other local authorities will be found with the documents on deposit.

Since the Northamptonshire plans have been transferred to the record office, most of the sealed plans from the main series have been removed from their wrappers, the seals broken, and the documents flattened. The unsealed plans have generally been left rolled in brown paper wrappers, although some of this wrapping was replaced (with the retention of the original labels) in 1998 during the examination of the documents for the preparation of this catalogue. The turnpike plans have also been left wrapped in brown paper with their original labels. The early canal plans, however, were removed from their wrapping in 1971, when negative photostats were made (now NRO, Maps 4057–4073) and the books of reference photocopied and the copies placed in a single box (NRO, X 5375). The maps themselves were repaired and in some cases the original labels have been pasted on to the backs of the maps, which are now kept rolled in individual map containers.

No register of plans deposited with the clerk of the peace survives among the records transferred to the county record office; the means of reference used in his office for both these and many other classes of record was a typescript list, probably dating from the mid twentieth century, arranged by subject, in which railway plans were briefly listed under the heading 'Railways', canal plans under 'Canals', and so on. A photocopy of this list is available in the record office index room and was, until the compilation of this catalogue, the only finding-aid for the plans listed here. It is therefore impossible to establish for certain whether any plans once deposited with the clerk had been lost before they were numbered in 1902.

The Soke of Peterborough deposited plans

In 1954 about a hundred plans and related documents deposited with the clerk of the peace of the Soke of Peterborough were transferred to the Northamptonshire Record Office.[1] Much of the material was in poor condition and in many cases the books of reference, plans and sections, and index maps had become separated from each other. Both the large-scale plans and the index maps, together with other maps, plans and architectural drawings transferred with the deposited plans, were catalogued into the NRO's main map series within the range Maps 1791–1911. Some of the books of reference were catalogued into the ML (i.e. Miscellaneous Ledger) series in the range ML 666–678; a few that were made up of large unbound sheets were treated as maps; and those that were still rolled inside the plans to which they related were left there. By no means all the books of reference contain *Gazette* notices. In a few cases two (or occasionally more) sets of documents relating to the same scheme survived, but only on a handful can any trace of a seal (long since broken) now be found and it is not clear whether second copies were regularly sealed in the Soke. A surprisingly large number of plans and books of reference lack a deposit memorial.

No register or other means of reference was transferred with the plans, none of which is itself numbered. Loose inside a few of the plans are what appear to be contemporary labels containing a brief note of their contents, but without any numbers.

For this catalogue, all the material in ML 666–678 and Maps 1791–1911 was re-examined and the items which can be identified as deposited plans have been included here in a list arranged in order of date of deposit. In the absence of a register it is impossible to decide

---

[1] P.I. King, 'Soke of Peterborough records in the Northamptonshire Record Office', *Peterborough's Past: The Journal of the Peterborough Museum Society*, i (1982–3), 35–40.

whether the plans now in archival custody include at least one copy of all those deposited in the Soke, although comparison with the better preserved Northamptonshire series (which, as already noted, includes plans for schemes which lay entirely within the Soke) suggests that the Soke series, as reconstructed here, is largely complete up to about 1880, but not thereafter.

Since their transfer to the record office, some of the Soke plans have been repaired but most have not. Except in the case of the one plan in the Soke series which has no Northamptonshire counterpart (and relates to a scheme whose limit of deviation came no nearer to any part of the county than the middle of the river Nene at the point where it forms the boundary between the Huntingdonshire parish of Fletton, in which the railway was to terminate, and the city of Peterborough), there should be no need to consult any of the Soke material, since the same documents, in better condition, can be found in the Northamptonshire series. In view of this duplication, and the fact that the Northamptonshire series also includes plans for schemes which lay entirely within Northampton, it should also be unnecessary to consult any of the plans deposited with the Borough.

Plans deposited with the clerk of the peace (or the clerk of the county council) of the Soke which were not transferred to Northampton in 1954 are at the Huntingdon branch of the Cambridgeshire Record Office. This material has not been examined during the preparation of this catalogue.

The arrangement of this catalogue

Most of this catalogue consists of a detailed list of the contents of what has been called the Main Series of deposited plans, i.e. the sets of documents numbered 1–250 by the clerk of the peace for Northamptonshire in 1902, and from 251 to 318 between then and 1960. In each case the title of the scheme has been transcribed from (normally) the plan cover, together with the name of the engineer and, where given, the surveyor. The separate profession of a railway surveyor, although briefly of great importance, was short-lived, since once the main network had been completed by about 1870 their work was absorbed into that undertaken by civil engineers who designed railways.[1] One firm of surveyors represented in the Northamptonshire plans, that founded by Charles Cheffins, continued as lithographers of railway plans and their name appears as such on several of the plans listed here. Lithographers' names, however, have not been included in the catalogue entries. The date of deposit has been noted but not the hour of the day, apart from the handful deposited in the early hours of 1 December 1845, rather than the previous day.

Except in a few cases where the set of documents contains extra items, the detailed contents have not been noted. Users may assume that all the documents required by the standing orders as they stood at the time of deposit are present, including (for most of the railway plans) an index map, plans and sections, book of reference and *Gazette* notice. Nor has the survival of both the sealed and unsealed copies of the plans been noted.

An attempt has been made to trace the fate of each project represented by a deposited plan and a note of the resulting Act included in the entry for schemes that were successful.[2] It would be possible (but has not been attempted here) to pursue the unsuccessful projects for which bills were introduced up to the point at which they were abandoned, using the *Journals* of the two Houses of Parliament and the relevant committee minutes at the House of Lords

---

[1] Biddle, *Railway Surveyors*, 82, 108–9.

[2] For the Acts see *Chronological Table of Local Legislation, 1797–1994* (1996); R. Devine, *Index to the Local and Personal Acts, 1797–1849* (1999); idem, *Index to the Local and Personal Acts, 1850–1995* (1996).

Record Office. For both these schemes, and those which never got to the stage of a bill being presented to Parliament, the local press would probably provide further information.

For all the Northamptonshire plans, the names of parishes through which the proposed scheme was to run have been given in the catalogue entry. Except for canals, turnpike roads and early railways promoted before the deposit of a *Gazette* notice was required, these have generally been taken from the latter source, rather than the plan or book of reference. The spelling of parish names has been modernised and, as far as possible, names of places that are not civil parishes removed. It is therefore possible to identify all the schemes that affected, however slightly, any parish, either in Northamptonshire or elsewhere.

A summary list of the Main Series plans has been inserted before the detailed entries, to make it easier to locate plans relating to a particular scheme. Immediately before this details are given of the subsidiary series of canal and turnpike plans deposited with the Northamptonshire clerk of the peace.

After the catalogue of the Northamptonshire plans there is a summary list of those deposited with the clerk of the peace for the Soke of Peterborough now in the record office, assembled from the material transferred in 1954, with cross-references to the corresponding Northamptonshire plan, which should be ordered in preference to the Soke copy, where both are available.

All the material described in this catalogue is available for inspection at the Northamptonshire Record Office without appointment during normal opening hours. None is suitable for photocopying but most items can, by arrangement, be photographed.

The value of deposited plans

The importance of deposited plans for the history of canals and railways (and in maritime counties docks and harbours) is obvious. They are basic to an understanding of how both systems evolved in a particular county, including abortive projects as well as those that were carried through.[1] In the case of the latter, especially the earlier canals promoted in the 1790s, it is possible to see how the final execution of a scheme differed from the original concept. At the other end of the canal era, the 1820s, it is possible to see glimpses of ambitious schemes to improve the line from London to the Midlands, through Northamptonshire, which came too late to escape being overtaken by the first viable proposals for a railway on the same route. During the Railway Mania of the 1840s, and to a lesser extent at other periods (the mid 1830s and the mid 1860s, for example), many projects which progressed no further are only properly recorded through the plans and other documents deposited by hopeful promoters, since contemporary newspaper accounts rarely include maps of the intended route. Similarly, for later tramway schemes, it is interesting to compare what was originally projected with how much was finally built.

Many of the later railway plans appear from their titles to be of very limited interest, even to railway historians, but it is worth remembering that the building of a new siding, or a minor deviation of a running line, might involve preparing a plan showing the existing layout at a particular station in considerable detail, as several plans deposited by the London & North Western Railway for both their Northampton stations and by the Great Northern for schemes at Peterborough well illustrate. Railway plans may also show lineside industry in

---

[1] For general accounts of Northamptonshire canals and railways see C. Hadfield, *The Canals of the East Midlands (including part of London)* (1966); R. Leleux, *The East Midlands* (Regional History of the Railways of Great Britain, ix) (1976).

detail at an earlier date than the Ordnance Survey.

For turnpike roads the deposited plans are less useful, since such schemes were only brought within the scope of the standing orders in 1807, when most of the network had been completed, but in Northamptonshire, as in other counties, there are a small number of plans of the last few improvements to be carried out by turnpike trusts, including a scheme of 1833 to build a paved way alongside Watling Street for steam-powered road vehicles. Similarly, the history of navigation on the river Nene goes back long before 1792, but there are several nineteenth-century plans relating to both navigation and drainage on the river.

The later nineteenth-century and twentieth-century plans for gas, water and electricity schemes have attracted little attention from local historians, which is a pity, since the provision of all three services is fundamental to the improvement of living standards, first in towns and more recently in the countryside. In a largely rural county such as Northamptonshire the extension of mains water and electricity to the countryside is arguably one of the reasons for the social transformation of the county's villages since the 1960s. The deposited plans are an important source (together with the records of the undertakings themselves) for tracing and dating the gradual spread of all three utilities and identifying the companies initially responsible for their provision.

The greatest neglect of the deposited plans, however, has come from local historians with no interest in transport history, much less in gas, water or electricity. Many people pursuing the general history of their community, or the topography of a particular area, or the origins of an individual house, fail to realise that deposited plans may provide additional information not available from other map sources. Admittedly, until the publication of this catalogue, it was difficult to use the Northamptonshire plans for general enquiries but by using the index here it is possible to identify every map which includes even a small part of a parish.

It will now be much easier for local historians in Northamptonshire to examine the relationship between their community and the railway in the nineteenth century. As well as establishing when the railway reached a particular village, and perhaps the original layout of the local station, it is possible also to see what earlier, unexecuted schemes there were for lines serving the same place, or later plans that never came to fruition. In this respect, deposited plans dating from before the publication of the large-scale Ordnance Survey maps in the early 1880s are especially valuable. Although the county is relatively well provided with eighteenth- or early nineteenth-century estate maps and inclosure plans, few parishes were subject to full-scale surveys under the Tithe Act of 1836, since most tithes were commuted at inclosure. There is therefore a dearth of maps showing the ownership, occupation and use of land in the 1840s, but for some parishes the gap is partly filled by the earlier railway plans, since these were drawn on a scale comparable to that of the tithe maps and the accompanying books of reference provide information similar to that found in the tithe apportionments. The large-scale inset plans required for built-up areas are helpful in showing village centres and portions of towns in great detail, especially as (unlike Ordnance Survey maps) they are accompanied by books of reference. Enlarged plans of a different kind were also sometimes provided to show the division of strips in open fields in parishes where these survived into the early railway age. Even after the Ordnance Survey becomes available in the 1880s, the schedules of owners and occupiers still provide useful additional information, for example about streets in Northampton affected by the extension of the tramway system.

The early canal and turnpike plans were drawn on a smaller scale than the railway plans but nonetheless still show the general layout of communities, as well as particular buildings. Canal plans tend to be especially careful in marking watermills, since millers were very concerned to ensure that such projects did not interfere with their power supply.

A final point, which adds to the value of the deposited plans compared with other maps of the same period, is that we know, with unusual precision, when they were prepared. They had to be deposited on 30 November in the year prior to the proposed application to Parliament; few were handed in before the deadline and some, prepared at the height of the Railway Mania, look as if they had been hurriedly completed only hours before they were deposited. Particularly in arable areas, such as Northamptonshire, it was difficult for surveyors to get access to the proposed route (even assuming the owners were cooperative, which some were not) until after the harvest had been got in. Therefore, we can be fairly certain that most of the earlier deposited plans for new lines (whether built or not) were surveyed and drawn in the autumn of the year in which they were deposited, and can probably be dated to within a few weeks, which is not the case with the first edition of the large-scale Ordnance Survey, where the surveys for a particular sheet may have been spread over a couple of years. Furthermore, there was great pressure on both the surveyors and those who collected the information for the book of reference to achieve absolute accuracy, since the documents deposited with the Private Bill Office would be scrutinised both by the committee examining the bill, to ensure that standing orders had been complied with, and by the scheme's opponents, in the hope of securing the rejection of the bill through failure to comply with standing orders. It would therefore have been unwise for surveyors simply to borrow the most recent estate map from a sympathetic landowner and sketch in any changes they could see on the ground, since the result might well be shown to be out of date or inaccurate, whereas some tithe maps give the impression of having been compiled in this way.

Deposited plans are a category of map source for local history that deserves to be better known and more widely used. One way of achieving that end is the compilation and publication of detailed catalogues, county by county, drawing the attention of all potential users, not merely railway historians, to the wealth of information available from the material. This publication will hopefully make the Northamptonshire plans easier to use and may perhaps encourage similar ventures in other counties.

NORTHAMPTONSHIRE

CANAL DEPOSITED PLANS

1. Plan of the intended navigable canal from Ashby de la Zouch, to join and communicate with the Coventry Canal at or near Griff, with the proposed navigable cuts or branches from Ashby de la Zouch to or near the lime works at Ticknal; lime works, lead mines and coal mines at Staunton Harold; lime works at Cloudhill and coal mines at Coleorton. Surveyed in 1792 by Robert Whitworth, engineer, John Smith, surveyor. No deposit memorial. Act 34 Geo. III c. xciii (1794).

Includes a table of distances and list of six 'Advantages of this Canal'. Branches from Cloud Hill, Staunton Harold and Ticknall join near Ashby de la Zouch, from where the main line runs to a junction with the Coventry Canal at Griff. No part of the route is in Northants.

2. A plan of the proposed Union Canal, to join the Leicester Navigation with a branch of the Grand Junction Canal, near Northampton; also of a branch to Market Harborough. Surveyed in 1792 by John Varley sen. and Christopher Staveley jun. Two copies. No deposit memorial. Act 33 Geo. III c. xcviii (1793).

From Leicester to Northampton, with a branch from near Foxton to Market Harborough. Includes table of distances.

3. A plan of the intended Union Canal from Leicester to join a branch of the Grand Junction Canal in the parish of [Blank] in the county of Northampton with two collateral branches one to Market Harborough the other to communicate with the Northampton Navigation. Surveyed in 1792 by John Varley sen. and Christopher Stavely (*sic*) jun. Deposited 11 Nov. 1792. The same scheme as **2**.

4. Plan of the intended Union Canal from Leicester to join a branch of the Grand Junction Canal, near Northampton; with a collateral branch to Market Harborough. Surveyed by John Varley and Christopher Staveley jun. in 1792. The same scheme as **2**.

Plan and book of reference. No deposit memorial on plan. Plan and book of reference both endorsed: 'No. 3. In obedience to an Act passed in the present session of Parliament intituled "An Act for making and maintaining a navigation from the town of Leicester to communicate with the river Nen, in or near the town of Northampton; and also a certain collateral cut from the said navigation"; I do this fourteenth day of May one thousand seven hundred and ninety-three hereby certify this map or plan to be one of the maps or plans required to be certified by me by the said Act. (Signed) Henry Addington, Speaker'.

Main Line: From Leicester, through Branston, Aylestone, Glen Parva, Wigston Magna, Newton Harcourt, Glen Magna, Burton Overy, Kibworth Harcourt, Kibworth Beauchamp, Fleckney, Saddington, Smeeton Westerby, Gumley, Foxton, Lubenham (Leics.); Hothorpe, Marston Trussell, Clipston, East Farndon, Great Oxendon, Kelmarsh, Maidwell, Lamport, Hanging Houghton, Cottesbrooke, Great Creaton, Spratton, Chapel Brampton, Dallington, Northampton and Duston (Northants.).

Branch to Market Harborough: In Lubenham and Great Bowden.

5. Plan of a proposed navigable cut from Buckingham to join the collateral cut from the Grand Junction Canal at Old Stratford. No surveyor named. Deposited 11 Nov. 1793. Act 34 Geo. III c. xxiv (1794).

From Buckingham, through Maids Moreton, Thornborough, Foscott, Leckhampstead, Thornton, Beachampton, Calverton (Bucks.); Wicken, Passenham and Old Stratford (Northants.).

6. A plan of the proposed line of navigable canal from Warwick to Braunston. Surveyed by James Sherriff, 1793. Deposited 11 Nov. 1793. Act 34 Geo. III c. xxxviii (1794).

From a junction with the Warwick & Birmingham Canal in Budbrooke, through Warwick, Radford Semele, Long Itchington, Birdingbury, Leamington Hastings, Grandborough, Woolscott, Willoughby (Warwicks.); terminating in a junction with the Oxford Canal in Braunston (Northants.).

7. A plan of the proposed canal from the Oxford Canal at Braunston in the county of Northampton, to join the river Thames at New Brentford in the county of Middlesex, to be called the Grand Junction Canal, with the collateral cuts or branches from the said canal to Daventry, Northampton, & Old Stratford, in the county of Northampton, and to Watford in the county of Hertford. No surveyor named. Act 33 Geo. III c. lxxx (1793).

Two copies; also a section. No deposit memorial on any of the three. Also a single sheet containing printed *Observations*, setting out benefits of the scheme, and an engraved 'Plan shewing the lines of the intended Grand Junction and Hampton Gay canals and their connections with the ports of London, Liverpool, Hull, Lynn & Bristol by means of the present inland canals' (undated). See also **18**.

Northamptonshire: Main Line: From Braunston, through Daventry, Welton, Thrup (Oxon.); Long Buckby, Whilton, Norton, Brockhall, Dodford, Weedon Beck, Stowe Nine Churches, Nether Heyford, Bugbrooke, Gayton, Blisworth, Stoke Bruerne, Roade, Grafton Regis, Yardley Gobion, Furtho, Cosgrove. *Branch to Daventry:* In Daventry and Thrup. *Branch to Northampton:* In Gayton, Rothersthorpe, Wootton, Hardingstone, Duston and Northampton. *Branch to Old Stratford:* In Cosgrove.

Buckinghamshire: Main Line: In Wolverton, Bradwell, Stantonbury, Great Linford, Newport Pagnell, Willen, Little Woolston, Great Woolstone, Woughton on the Green, Fenny Stratford, Water Eaton, Stoke Hammond, Soulbury, Linslade.

Bedfordshire: Main Line: Leighton Buzzard (Beds.); Grove, Slapton, Ivinghoe, Pitstone, Cheddington, Marsworth (Bucks.).

Hertfordshire: Main Line: Tring, Aldbury, Northchurch, Berkhampstead, Bovingdon, Hemel Hempstead, Abbots Langley, Kings Langley, Watford, Rickmansworth. *Branch to Watford:* Rickmansworth, Watford.

Buckinghamshire: Main Line: Denham.

Middlesex: Main Line: Harefield, Uxbridge, Cowley,

Hillingdon, Drayton, Harlington, Hayes, Southall, Heston & Isleworth, Norwood, Hanwell, New Brentford.

8. A plan of that part of the open fields of Braunston, through which the Oxford Navigable Canal is lately cut. No surveyor named. No deposit memorial. Undated.

Marks and names individual furlongs in the fields, and houses and crofts in the village.

9. Listed by the clerk of the peace as 'Oxford Canal (Braunston to Brentford)', which appears to be a description of the Grand Junction Canal, as in **7** and **18**. Missing since at least 1971.

10. A map of the intended Grand Union Canal from the Leicestershire and Northamptonshire Union Canal in the parish of Gumley, in the county of Leicester, to the Grand Junction Canal in the parish of Norton near Buckby Wharf, in the county of Northampton. Surveyed by B. Bevan, 1808–9. Deposited 28 Sept. 1809. Act 50 Geo. III c. cxxii (1810). Usually known today as the Old Union Canal to distinguish it from the Grand Union Canal created in 1929.

Main Line: From Gumley, through Foxton, Lubenham, Theddingworth, Husbands Bosworth, North Kilworth (Leics.); Welford, Stanford, Elkington, Yelvertoft, Winwick, Crick, Watford, Welton, Norton (Northants.).

Branch to Welford Road Bridge: In Husbands Bosworth (Leics.); Welford (Northants.).

Reservoirs and Feeders and new channel to Brooks: In Husbands Bosworth (Leics.); Sulby, Welford, Naseby, Cold Ashby (Northants.).

Crick Reservoir: In Crick (Northants.).

11. Union Canal. Plan of deviation of intended line of canal. No surveyor named. Deposited 20 Sept. 1804. Act 45 Geo. III c. lxxi (1805).

Listed on the clerk of the peace's label as 'Branch in connection with the Leicestershire & Northants. Union Canal'. Shows a canal from the outskirts of Market Harborough through Foxton to Gumley Debdale Wharf.

Two books of reference:

(a) Union Canal. Book of reference to plan of intended deviation of line of canal marked (A). In Gumley, Foxton, Great Bowden, Lubenham (Leics.).

(b) A list of the land owners and tenants whose land the intended extension of the Leicestershire and Northamptonshire Union Canal is to pass through with the situations of such land from the bason of such canal in the parish of Gumley in the county of Leicester into a close or ground inclosed in the parish of Great Bowden in the said county of Leicester belonging to Edward Dawson of Long Whatton in the same county esqr. in the occupation of John Mutton, called Kingston Close, and through such close to where the turnpike road from Leicester to Market Harborough bounds the same so as to make a communication there with the said turnpike road; being a reference to the plan left herewith, marked with the letter A. In Gumley, Foxton, Great Bowden, Harringworth, Market Harborough (Leics.).

12. A map of the intended Grand Union Canal from the Leicestershire and Northamptonshire Union Canal in the parish of Foxton in the county of Leicester to the Grand Junction Canal in the parish of Braunston in the county of Northampton. Surveyed by B. Bevan, 1808. Deposited 29 Sept. 1808. No Act (superseded by **10**).

From the Union Canal in Foxton, through Lubenham, Theddingworth, Husbands Bosworth, North Kilworth (Leics.); Welford, Stanford, Elkington, Yelvertoft, Winwick, Crick, Kilsby, Barby, Willoughby, Braunston (Northants.); terminating in a junction with the Grand Junction Canal. Includes a navigable feeder in Husbands Bosworth and Welford.

13. A map of the intended canal from the Union Canal near Harborough in the county of Leicester to the Welland Navigation at Stamford in the county of Lincoln. Surveyed by B. Bevan, 1810. Deposited 29 Sept. 1810. No Act.

Book of Reference: From Hothorpe (Northants.), through Husbands Bosworth, Theddingworth, Lubenham, Foxton, Kibworth Harcourt, West Langton, East Langton, Great Bowden (Leics.); Sutton Bassell, Weston by Welland, Ashley, East Carlton, Middleton, Cottingham, Rockingham, Gretton, Harringworth, Wakerley, Duddington (Northants.); Tixover (Rutland); Collyweston, Easton, Wothorpe, Stamford Baron St Martin Without (Northants.).

Plan: Two feeders, one beginning in two reservoirs in Husbands Bosworth, through Theddingworth to join the Union Canal in Lubenham (Leics.), the other from a reservoir in West Langton, through East Langton to join the Union Canal in Great Bowden (Leics.); and a canal from a junction with the Union Canal in Great Bowden, through Sutton, Weston by Welland, Ashley, East Carlton, Middleton, Cottingham, Rockingham, Gretton, Harringworth, Wakerley, Duddington, Collyweston, Easton on the Hill and Wothorpe, terminating in a junction with the Welland Navigation in Stamford (Northants.).

14. A map of the intended canal and navigation from the Union Canal near Harborough in the county of Leicester to Stamford and Spalding in the county of Lincoln and Peterborough in the county of Northampton. Surveyed by B. Bevan, 1810. Deposited 29 Sept. 1810. No Act.

Plan: Follows the same course (with the same feeders) as **13** as far as Stamford, from where it continues through Uffington, Tallington, West Deeping, Market Deeping, Deeping Gate, Deeping St James, Crowland, Spalding Common. A second sheet shows a canal from Peterborough, through Stanground, Eye, Borough Fen, Paston, Gunthorpe and Werrington; with another line passing through Crowland, Borough Fen, Glinton and Peakirk to join the first line shown on the second sheet.

Book of reference:

Main Line: From Hothorpe (Northants.); through Husbands Bosworth, Theddingworth, Lubenham, Foxton, Kibworth Harcourt, West Langton, East Langton, Great Bowden (Leics.); Sutton Basset, Weston by Welland, Ashley, East Carlton, Middleton, Cottingham, Rockingham, Gretton, Harringworth, Wakerley, Duddington (Northants.); Tixover (Rutland); Collyweston, Easton on the Hill, Wothorpe, Stamford Baron St Martin Without (Northants.); Stamford, Uffington, Tallington, West Deeping, Market Deeping, Maxey, Deeping Gate, Deeping St James, Crowland, Spalding (Lincs.).

Canal from Deeping to Peterborough: Through Maxey, Glinton, Peakirk, Werrington, Gunthorpe, Paston, Eye, Peterborough (Northants.); Stanground (Cambs. & Hunts.).

Feeeder to the canal from Peterborough to Deeping: Through Werrington and Paston (Northants.).

At end: List of the Adventurers of Deeping Fen.

15. Plan of the proposed Stamford Junction Navigation, from Oakham in the county of Rutland, to Stamford and Boston in the county of Lincoln, and from Stamford to Peterborough in the county of Northampton. Surveyed under the direction of Thomas Telford by Hamilton Fulton and drawn by W.A. Provis. 1810. Deposited 17 Sept. 1810. No Act.

Plan: From a junction with the Oakham Canal at Oakham, through Egleton, Martinsthorpe, Manton, Lyndon, North Luffenham, Ketton, Easton on the Hill, Wothorpe, Stamford and Uffington; from where one branch continues through Tallington, West Deeping, Market Deeping, Deeping St James, Peakirk, Werrington, Gunthorpe, Paston, Eye and Peterborough, terminating in Stanground in a junction with the river Nene; the other branch continues through Barholm, Greatford, Thurlby, Bourne, Morton, Hacconby, Dunsby, Rippingale, Dowsby and Aslackby, and from there following the South Forty Foot Drain to terminate in a junction with the river Witham at Boston. Enlarged plans of the built-up areas of Oakham, Peakirk, Tallington, Bourne, Thurlby, Stamford and Ketton.

Book of reference: From Oakham, through Martinsthorpe, Manton, Lyndon, North Luffenham, South Luffenham, Ketton, Tinwell (Rutland); Easton on the Hill, Wothorpe, Stamford Baron St Martin Without (Northants.); Stamford, Uffington, Tallington, Barholm, Greatford, Thurlby, Baston, Bourne, Morton, Hacconby, Dunsby, Rippingale, Dowsby, Aslackby, Pointon, West Deeping, Market Deeping (Lincs.); Maxey, Peakirk, Glinton, Paston, Werrington, Peterborough, Gunthorpe, Eye (Northants.); Stanground (Cambs.).

Reservoirs at Braunston and Brooke: In Braunston, Oakham, Brooke and Gunthorpe (Rutland).

Reservoir on the river Wash near Ryhall: In Ryhall (Rutland).

16. Nene Outfall. A plan of the line of certain intended alterations in the present channel or river from the city of Peterborough to the sea, for improving the navigation thereof and the outfall of the river Nene, and for the better draining the several lands discharging their waters thereby and of the several lands through or over which the said alterations and improvements are intended to be carried, or made. 18 Sept. 1822. No engineer or surveyor named. Deposited 30 Sept. 1822. No Act.

17. A plan and section of the river Nene from and above the city of Peterborough in the county of Northampton; to and below the town of Wisbech in the Isle of Ely and county of Cambridge. W. Swansborough, civil engineer, 27 Sept. 1823. Deposited 30 Sept. 1823. No Act.

18. Plan of the proposed canal leading from Braunston in the county of Northampton to join the river Thames at New Brentford in the county of Middlesex called the Grand Junction Canal with the collateral cuts or branches from the said canal to Daventry, Northampton and to Old Stratford in the county of Northampton, and to Watford in the county of Hertford. No engineer named. Deposited 10 Nov. 1792. Act 33 Geo. III c. lxxx (1793). See also **7**.

19. Plan and section of certain rivers or cuts called Hills Cut and Smith's Leam and of that part of the river Nene extending from Hills Cut to Goldiford Stanch and of that part of the Wisbech river extending from Smith's Leam to the limits of the port of Wisbech. W. Swansborough, [engineer], 1826. Deposited 11 Nov. 1826. Act 8 Geo. IV c. lxxxv (1827).

20. Plan and sections of the intended improvement of the outfall of the river Nene by a new cut or channel from Kinderley's Cut to Crabhole. W. Swansborough, [engineer], 1826. Deposited 19 Oct. 1826. Act 8 Geo. IV c. lxxxv (1827).

21. Plan of the intended London and Birmingham Junction Canal with collateral branch. Surveyed under the direction of Thomas Telford, civil engineer, 1827, by Dugdale Houghton, land surveyor. Deposited 30 Nov. 1827. No Act.

Main line: From a junction with the Stratford Canal in Solihull, through Tanworth, Hampton in Arden, Berkswell, Stoneleigh, Coventry, Stoke, Wyken, Walsgrave on Sowe, Brinklow, Kings Newnham, Monks Kirby, Newbold on Avon, Bilton, Rugby, Hillmorton, Willoughby (Warwicks.); Barby, Onley, Braunston (Northants.), terminating there in a junction with the Grand Junction Canal and Oxford Canal.

Branch: In Coventry, Stoke and Foleshill (Warwicks.).

22. Plan and section of proposed new drain from Clow's Cross to the new outfall near Buckworth Sluice. Thomas Peeir, [engineer], Nov. 1828. Deposited 28 Nov. 1828. Act 10 Geo. IV c. civ (1829).

23. Plan and section of the intended variation and extension of the Nene Outfall Cut from Sutton-Wash to Buckworth Sluice. W. Swansborough, [engineer], 1828. Deposited 28 Nov. 1828. Act 10 Geo. IV c. civ (1829).

24. London & Birmingham Canal. Surveyed under the direction of James Walker by Dugdale Houghton, Birmingham. Deposited 30 Nov. 1836. No Act.

Canal: From a junction with the Stratford on Avon Canal in Lapworth, through Rowington, Wroxall, Hatton, Shrewley, Haseley, Beausale, Warwick, Leek Wootton, Milverton, Lillington, Cubbington, Offchurch, Long Itchington, Bascote, Southam, Ladbroke, Lower Radbourne, Wormleighton, Stoneton, Priors Hardwick, Upper Radbourne (Warwicks.); Upper & Lower Boddington (Northants.); Claydon, Cropredy, Prescote, Clattercote, Wardington (Oxon.); Chipping Warden, Edgcott, Byfield, Culworth, Sulgrave, Helmdon, Weedon Beck, Wappenham, Slapton, Abthorpe, Bradden, Greens Norton, Towcester, Easton Neston, Shutlanger, Stoke Bruerne (Northants.), terminating there in a junction with the Grand Junction Canal.

Reservoir, feeders and brooks: In Stuchbury, Helmdon, Wappenham, Astwell with Falcutt, Slapton, Abthorpe, Bradden, Greens Norton and Towcester (Northants.)

25. Map of the intended improvements along part of the line of the existing Oxford Canal between Longford in the county of the city of Coventry and Wolfampcote in the county of Warwick (together with the sections thereof). Laid down from surveys and levels taken by, and under the immediate direction of, Charles Vignoles, engineer, 8 Oct. 1828. Deposited 28 Nov. 1828. Act 10 Geo. IV c. xlviii (1829).

Plan includes enlarged details of Foleshill, Walsgrave on Stowe, Shilton, Ansty, Brinklow, Stretton under Fosse, Newbold upon Avon, Brownsover, Hillmorton and Braunston.

Book of reference lists lands under three headings:

Land through which the proposed new cuts or canal are intended to be made: Foleshill, Walsgrave on Sowe, Shilton, Ansty, Withybrook, Stretton under Fosse, Monks Kirby, Easenhall, Kings Newnham, Church Lawford, Newbold upon Avon, Little Lawford, Brownsover, Clifton upon Dunsmore, Hillmorton (Warwicks.); Kilsby, Barby, Onley (Northants.); Willoughby (Warwicks.); Braunston (Northants.); Wolfhampcote (Warwicks.).

Land through which the parts of the existing canal intended to be retained pass: Foleshill, Walsgrave on Sowe, Shilton, Ansty, Withybrook, Stretton under Fosse, Monks Kirby, Brinklow, Easenhall, Kings Newnham, Church Lawford,

Newbold upon Avon, Little Lawford, Brownsover, Clifton upon Dunsmore, Hillmorton (Warwicks.); Kilsby, Barby, Onley (Northants.); Willoughby (Warwicks.); Braunston (Northants.); Wolfhampcote (Warwicks.).

Land through which the parts of the existing canal intended to be abandoned pass: Walsgrave on Sowe, Shilton, Ansty, Withybrook, Stretton under Fosse, Monks Kirby, Brinklow, Easenhall, Kings Newnham, Church Lawford, Newbold upon Avon, Little Lawford, Cosford, Brownsover, Clifton upon Dunsmore, Hillmorton (Warwicks.); Kilsby, Barby, Onley (Northants.); Willoughby (Warwicks.); Braunston (Northants.); Wolfhampcote (Warwicks.).

NORTHAMPTONSHIRE TURNPIKE ROAD DEPOSITED PLANS

1A. Map of an intended turnpike road leading from the Southam turnpike road in the parish of Leamington Hastings in the county of Warwick to the London turnpike road in the parish of Braunston in the county of Northampton. Thomas Hopcraft jun., [surveyor]. Deposited 29 Sept. 1809. No Act.

From Leamington Hastings, through Grandborough and Wolfhampcote (Warwicks.) to Braunston (Northants.).

2A. A map or plan of that part of Watling Street Road which passes by, through or near the parishes of Aston-Flamville, Claybrook, Bitteswell, Lutterworth and Cottesbach in the county of Leicester, Stretton-Baskerville, Burton-Hastings, Wolvey , Wibtoft (in the parish of Claybrook), Willey, Monkskirby & Churchover in the county of Warwick, Crick, Kilsby, Watford, Ashby-Legers, Buckby, Welton, Norton (by Daventry), Whilton, Brockhall, Flore, Dodford, Upper Heyford, Nether Heyford and Weedon in the county of Northampton, which is not now turnpike, but for which application is intended to be made in the next session of Parliament for the usual powers and authorities for making the same turnpike. No surveyor named. Received 28 Sept. 1809. No Act. Route as described in title.

3A. A plan of the road leading from the north corner of the Sessions House in the town of Buckingham, to join the turnpike road leading from the town of Buckingham, in the county of Bucks., to the north extent of the parish of Hanwell, in the county of Oxford. 1810. No surveyor named. Deposited 29 Sept. 1810. Act 51 Geo. III c. ii (1811).

Plan shows section of the town of Buckingham as in title; no part of the route is in Northants. Attached is a printed handbill announcing the sale of a freehold estate in Welby, consisting of a farmhouse and buildings, public house, cottages and 115 acres of land in the occupation of Mr Chaloner, to be sold by auction at the Hind Inn, Wellingborough, 13 June 1810. This appears to have been used as a wrapper for the plan. Also enclosed in the modern wrapper is a printed table of fees to be taken by the clerks to the justices of the peace in Northants. (undated, but said to supersede a table made in 1789 and confirmed at the Lent Assize, 1790). This may also have served as a wrapper for the plan.

4A. Plan of the proposed road, to lead from the Sessions House in Buckingham, to Newport Pagnell, in the said county. 1814. No surveyor named. Received 29 Sept. Act 55 Geo. III c. lxxv (1815).

From Buckingham, through Maids Moreton, Foscott, Leckhampstead (Bucks.); Wicken, Passenham (Northants.); Wolverton, Bradwell, Stantonbury, Great Linford and Newport Pagnell (Bucks.).

5A. Map of the intended turnpike road from Kettering in the county of Northampton, to the town of Northampton in the said county, leading from the east end of a certain lane, called Hall Lane, in the parish of Kettering, through the several parishes and townships of Kettering, Broughton, Pitchley, Walgrave, Orlinbury, Hannington, Sywell, Overstone, Moulton, Weston Favell, Abington and Kingsthorpe, and the parish of St Giles, in the town of Northampton, to the east end of Abington Street, in the said town. Surveyed by T. Lilburne, 1818. Received 30 Sept. 1818. Act 59 Geo. III c. xli (1819).

Route as in title. Plan and dummy book of reference, containing a note that it is not intended that the existing road should be altered to take in any other land, or deviate from the line described in the plan, without the consent of the owners of the land in question.

6A. Wrapper containing two deposits:
(a) **Plan of the present road from the Towcester & Oxford road to Buckingham proposed to be converted into a turnpike road. 25 Sept. 1822.** No surveyor named. Deposited 30 Sept. 1822. No Act.
(b) **Plan of the intended turnpike road from the town of Buckingham to the Oxford and Northampton turnpike road in the open fields of Silston.** Surveyed by John King, 1823. Deposited 27 Sept. 1823. Act 5 Geo. IV c. cxli (1824).

From Buckingham, through Maids Moreton, Akeley, Lillingstone Dayrell (Bucks.), Lillingstone Lovell (Oxon.) and Whittlebury to Silverstone (Northants.).

7A. Plan of the proposed turnpike road from Northampton to Bedford. 1826. [John?] Durham, Dunstable, surveyor. Deposited 23 Nov. 1826. Act 7 & 8 Geo. IV c. lxxi (1827).

From St Giles, Northampton, through Hardingstone, Great Houghton, Little Houghton, Brafield on the Green, Denton, Castle Ashby, Yardley Hastings (Northants.), Warrington, Lavendon and Cold Brayfield (Bucks.).

8A. Plan for improving the Geese Bridge Valley on the Holyhead Road. John Macneill, resident engineer to the Parliamentary Commissioners of the Holyhead and Liverpool Road. Deposited 15 Nov. 1830. No Act.

In Nether Hayford, Bugbrooke, Stowe Nine Churches, Pattishall and Cold Higham (Northants.).

8B. Map shewing that part of the Holyhead Road which lies between the town of Birmingham and the city of London, of which ten feet in the width of the carriage way or waste ground on the side of the carriage way, is intended to be constructed or improved into a hard and solid road for the passing or travelling thereon of locomotive steam engines. John Macneill, civil engineer, 28 Nov. 1833. No Act.

From Birmingham, through Aston (Warwicks.); Yardley (Worcs.); Bickenhill, Sheldon, Elmdon, Hampton in Arden, Packington, Meriden, Allesley, Coventry, Stretton, Thurlaston, Dunchurch, Woolscott, Willoughby (Warwicks.), Braunston, Daventry, Newnham, Dodford, Weedon Beck, Stowe Nine Churches, Nether Heyford, Bugbrooke, Cold Higham, Pattishall, Greens Norton, Towcester, Paulerspury, Whittlebury, Potterspury, Cosgrove, Furtho, Passenham (Northants.); Wolverton,

Stony Stratford, Calverton, Bradwell Abbey, Loughton, Shenley, Fenny Stratford, Little Brickhill, Great Brickhill (Bucks.); Potsgrove, Battlesden, Hockliffe, Chalgrove, Tilsworth, Houghton Regis, Dunstable, Kensworth, Caddington (Beds.); Flamstead, Harpenden, Redbourn, St Albans, Shenley, Ridge, East Barnet, Friern Barnet (Herts.); South Mimms, Hadley, Finchley, Hornsey and Islington (Middlesex).

9A. Warwick and Northampton Road. Plan of proposed alterations in the road through Staverton, Northamptonshire. 1831. C. Pixell, surveyor, Warwick. Deposited 30 Nov. 1831. Act 2 & 3 Wm IV c. xcviii (1832).

Large-scale plan of the village, showing a realignment of the road through the built-up area and a new road by-passing the village to the south.

10A. Number not used in the clerk of the peace's list; either never used or a plan has long been lost.

10B. A plan of the projected new branch road, leading out of the turnpike road at the bridle-gate at the foot of the hill at the western extremity of the village of Aynho in the county of Northampton to and through the village of Souldern in the county of Oxford into the turnpike road at Souldern turnpike-gate. John and Richard Davis, land surveyors, Banbury, 26 Nov. 1831. Deposited 28 Nov. 1831. Act 2 & 3 Wm IV c. xvi (1832).

In Aynho and Souldern, showing a branch road to the west of Aynho Park and through Souldern village.

11A. A map shewing a propos'd line line of road at Staverton, Northamptonshire, by R. Pettifer. 1831. Plan and book of reference. Deposited 9 June 1831. Similar to **9A** but showing only the by-pass to the south of the village.

12A. A plan of the projected turnpike road leading from Clifton turnpike through the village of Souldern in the county of Oxford to Souldern turnpike. 1833. Richard Davis, Banbury, [surveyor]. Plan and book of reference. Deposited 30 Nov. 1833. No Act.

A revised version of the scheme shown in **10B** in Aynho and Souldern.

13A. Map or plan of an intended branch turnpike road from near the village of Badby to Dodford Lane End, in the parish of Newnham in the county of Northampton. Richard Davis, land surveyor, Banbury. Deposited 30 Nov. 1838. No Act.

In Badby and Newnham. The book of reference describes the scheme as a projected branch of the turnpike road leading from Banbury to Lutterworth.

14A. Plan of the road commencing near the west end of Thrapston Bridge in the parish of Islip in the county of Northampton and terminating in the parish of Stanion in the said county in the turnpike road leading from Kettering to Stamford. Robert Russel, surveyor, Brackley, Nov. 1839. Deposited 30 Nov. 1839. No Act.

Plan, two copies of the book of reference, print of House of Lords standing orders (1839), and sealed duplicate. Through Islip, Aldwincle, Lowick, Sudborough, Brigstock and Stanion (Northants.). Enlarged plans of Islip, Lowick, Brigstock and Stanion villages.

15A. Map or plan and section of an intended branch turnpike road from the Banbury and Lutterworth turnpike road at or near Badby Bridge to the Stratford and Dunchurch turnpike road at or near Dodford Lane. John Durham jun., [surveyor], 30 Nov. 1839. Deposited 30 Nov. 1839. Act 3 & 4 Vict. c. xxxviii (1840).

Plan, two copies of the book of reference, print of House of Lords standing orders (1839), and sealed duplicate. In Badby and Newnham (Northants.); detailed plan of Newnham village.

NORTHAMPTONSHIRE MAIN SERIES PLANS

SUMMARY

1	London & Birmingham Railway	1830
2	London & Birmingham Railway	1831
3	Plan of a Proposed Rail-Way from London to Birmingham	1832
4	London & Birmingham Railway	1834
5	London & Birmingham Railway. Deviation of line of railway from Watford (Northants.) to Hillmorton (Warwicks.)	1835
6	Northern & Eastern Railway. Plan and section of the extended line from Cambridge to York	1836
7	North & South Junction Railway	1836
8	South Midland Counties Railway, with branches to Leicester and Stamford	1836
9	South Midland Counties Railway, with branches to Leicester and Stamford. Identical to 8.	1837
10	Railway commencing at a junction with the London & Birmingham Railway in Gayton; terminating near Peterborough	1842
10a	River Nene Improvements &c.	1840
11	Oxford & Rugby Railway	1844
12	London & York Railway. Alternative line [to that shown in 13 but numbered first] commencing at 76 miles 5 furlongs and terminating at 82 miles 5 furlongs from Old St Pancras Road at King's Cross, London	1844
13	Direct Northern Railway from London to York by way of Lincoln	1844
14	London & Worcester & Rugby & Oxford Railway with an extension from Worcester to Dudley [and Wolverhampton *crossed through*]	1844
15	Midland Railways (*sic*) Extensions. Syston & Peterborough Railway	1844
16	Cambridge & Lincoln Railway	1844
17	London & York Railway. Alternative line commencing at 76 miles 5 furlongs and terminating at 82 miles 5 furlongs from Old St Pancras Road, King's Cross	1844
18	London, Birmingham & Buckinghamshire Railway	1845
19	Boston, Stamford & Birmingham Railway, from Birmingham to Boston and Wisbech, with a branch to Market Harborough	1845
20	Leicester & Bedford Railway	1845
21	London & Birmingham Railway. Banbury Line	1845
22	Warwickshire & London Railway. Plan and Section from Worcester to Weedon	1845
23	Direct London & Manchester Railway, with a branch to Crewe	1845
24	Direct Northern Railway from London to York, with branches	1845
25	Northampton, Bedford & Cambridge Railway	1845
26	Sheffield, Nottingham & London Direct Railway. Line of Railway from Nottingham to Chesterfield	1845

27	London & York (Great Northern) Railway. Stamford & Spalding Branch	1845
28	Stamford, Market Harborough & Rugby Railway, with a branch to the Leicester & Bedford Railway	1845
29	Midland & Eastern Counties Railway	1845
30	South Midlands & Southampton Junction Railway, from Northampton to Reading and Basingstoke	1845
31	Northampton, Lincoln & Hull Direct Railway	1845
32	Oxford, Witney, Cheltenham & Gloucester Independent Extension Railway, with two branches therefrom	1845
33	Rugby & Huntingdon Railway, with a branch by Market Harborough to Stamford	1845
34	East & West of England Junction Railway	1845
35	Peterborough, Spalding & Boston Junction Railway	1845
36	London, Warwick, Leamington & Kidderminster Railway, with branch to the Birmingham & Gloucester Railway at Coston (sic) Hackett	1845
37	Rugby & Stamford Railway	1845
38	Buckinghamshire Railway. Tring and Banbury Line	1845
39	Midland & Eastern Counties Railway	1845
40	Peterborough & Nottingham Junction Railway	1845
41	Midland Railway. Branches and Deviations from the Syston & Peterborough Railway	1845
42	South Midland, or Leicester, Northampton, Bedford & Huntingdon Railways. Plan and section endorsed with memorial dated 30 Nov. 1845 (signed by Charles Markham); book of reference (which corresponds with plan) bears printed date Nov. 1846 and memoriam (signed by H.P. Markham) dated 30 Nov. 1846	1845/6
43	London & Nottingham Railway	1845
44	Port of Wisbech, Peterborough, Birmingham & Midland Counties Union Railway	1845
45	Lynn, Wisbeach & Peterborough, Midland Counties & Birmingham Junction Railway	1845
46	Peterborough, Wisbech & Lynn Junction Railway	1845
47	Northampton, Banbury & Cheltenham Railway	1845
48	London & Birmingham Railway. Extensions at Birmingham; at Leamington; from Weedon to Northampton; at Coventry; at Rugby	1845
49	Eastern Counties Railway. York Extension. Line between Cambridge and Lincoln, with branches to Sleaford, Boston and Spalding	1845
50	London & Birmingham Extension, Northampton, Daventry, Leamington & Warwick Railway, with branches to Rugby and Fenny Compton	1845
51	Midland Grand Junction Railway; extending from the town of Reading, Berkshire, to the station at Blisworth, Northamptonshire	1845
52	Great Northern Railway. Deviations between London and Grantham	1846
53	Buckingham & Brackley and Oxford & Bletchley Junction Railway. Extensions to Banbury and Aylesbury and into Oxford	1846
54	Boston, Stamford & Birmingham Railway. Peterborough & Thorney Line	1846
55	Northampton & Banbury Railway	1846
56	Midland Railway Extensions. From Leicester to Hitchin and to Northampton and Huntingdon	1845

57	Drainage of Crowland Cowbit Washes and Other Lands	1846
58	Midland Railway. Syston & Peterborough Railway. Deviations and approach to Manton Station	1846
59	Great Northern Railway. Deviations between Peterborough, Doncaster and Boston	1846
60	Eastern Counties Railway. Peterborough to Folkingham	1846
61	London & North Western Railway. Newport Pagnell, Olney Branch	1846
62	River Nene. Improvement and Drainage	1847
63	Midland Railway. Alteration of the line and branches near Wellingborough and station approaches	1847
64	Nene Valley Drainage and Navigation Improvement	1851
65	Stamford & Essendine Railway, with a branch to join the Midland Railway at Stamford, and works for the improvement of the river Welland at or near Stamford	1852
66	London & North Western Railway. Branch from Northampton to Market Harborough	1852
67	Midland Railway. Extension from near Leicester via Bedford to Hitchin, and branches; and approaches to the intended station at Wellingborough	1852
68	South Wales & Northamptonshire Junction Railway.	1853
69	Stamford & Essendine Railway. Additional lands required for station and other purposes	1856
70	London & North Western Railway. Additional Works	1857
71	Northampton Gas Light Co. Additional Land	1857
72	Nene Valley Drainage & Navigation Improvements	1858
73	London & North Western Railway. Denton to Stalybridge [etc.]	1858
74	Weedon & Leamington Railway	1858
75	Northampton Waterworks	1860
76	Stamford & Essendine Railway	1861
77	London & North Western Railway. Additional Powers	1861
78	Daventry Railway	1861
79	Eastern Counties Railway. Railways between Wisbech and Peterborough	1861
80	Kettering & Thrapston Railway	1861
81	Peterborough, Wisbech & Sutton Railway	1862
82	London & North Western Railway. Additional Powers	1862
83	Midland Railway. New Lines and Additional Powers	1862
84	Northampton & Banbury Junction Railway	1862
85	East & West Junction Railway. Blisworth to Worcester, with branches	1862
86	Market Harborough & Melton Mowbray Railway	1862
87	Kettering & Thrapstone Railway. Huntingdon Extension	1862
88	Market Harborough & East Norton Railway	1863
89	Great Eastern Northern Junction Railway	1863
90	Stamford & Essendine Railway. Sibson Extension	1863
91	Market Harborough & Melton Mowbray Railway	1863
92	Blockley & Banbury Railway	1863

93	Daventry Railway. Extensions to Southam and Leamington	1863
94	East & West Junction Railway	1863
94a	Chipping Norton & Banbury Railway	1863
95	Northampton & Banbury Railway. Deviations &c.	1864
96	Banbury Water	1864
97	Lancashire & Yorkshire and Great Eastern Junction Railway	1864
98	Bourton, Chipping Norton & Banbury Railway	1864
99	Bedford Northampton & Weedon Railway	1864
100	Peterborough Wisbech & Sutton Railway. Extensions and Deviations	1864
101	Chipping Norton, Banbury & East & West Junction Railway	1864
102	Northampton & Banbury Junction Railway. Extensions	1864
103	Bedford Northampton & Leamington Railway	1864
104	East & West Junction Railway. Extension	1864
105	Chipping Norton, Banbury, East & West Junction Railway	1865
106	Bedford & Northampton Railway. Deviations &c.	1865
107	East & West Junction Railway. Hitchin Extension	1865
108	Newport Pagnell Railway. Extensions to Bedford & Northampton, Northampton & Peterborough and Leicester & Hitchin Railways	1867
109	Peterborough Waterworks	1865
110	Bedford & Northampton Railway	1866
111	Peterborough Waterworks	1866
112	Weedon & Daventry Railway	1867
113	Peterborough Gas	1867
114	Wellingborough Waterworks	1867
115	Nene Valley Drainage and Navigation Improvements	1861
116	Midland Railway. Additional Powers	1867
117	Thrapston Market	1869
118	London & North Western Railway. Additional Powers	1868
119	Newport Pagnell Railway. Alteration of Levels	1869
120	Midland Counties & South Wales Railway. Extension to Buckinghamshire Railway	1869
120a	Northampton Cattle Market	1869
121	Northampton Gas Light Co.	1870
122	Northampton Extension and Improvement. Plans and sections of proposed street improvements and sewage outfall and irrigation works	1870
123	Midland Railway. Additional Powers	1870
124	Weedon & Northampton Junction Railway	1870
125	Northampton & Banbury Junction Railway. Railway from Banbury to Blockley. Deviation in Authorised Extension to Ross	1870

126	Great Western Railway. Substituted Railway, New Railways, Roads, Additional Lands &c	1870
127	Coal Owners' Associated London Railway	1870
128	Towcester & Hitchin Railway	1871
129	Northampton & Banbury Junction Railway. Extension to Northampton	1871
130	Daventry & Weedon Railway	1871
131	Kettering Waterworks	1871
132	Midland Railway. Nottingham and Rushton Lines	1871
133	Northampton & Daventry Junction Railway	1871
134	East & West Junction Railway	1872
135	Midland and Manchester, Sheffield & Lincolnshire Railways. Extensions	1872
136	Midland Railway. Additional Powers	1872
137	Market Harborough, Melton Mowbray & Nottingham Railways	1872
138	Banbury & Cheltenham Direct Railway	1872
139	Great Northern Railway. Additional Powers	1872
140	Blockley & Banbury Railway	1872
141	London & North Western Railway. New Lines	1872
142	Great Northern and London & North Western Railways. New Lines from Market Harborough and Stathern to Bingham	1873
143	Peterborough Gas Works. Deposited 29 Nov	1873
144	London & North Western Railway. England and Ireland	1873
145	Midland Railway. Additional Powers. Kettering & Manton Line	1873
146	Great Northern Railway. Further Powers	1873
147	London & North Western Railway. New Lines and Additional Powers	1874
148	East & West Junction Railway. Blisworth Branch	1874
149	London & North Western Railway. Bletchley, Northampton and Rugby Railway	1874
150	Peterborough Water Co.	1874
151	Buckinghamshire & Northamptonshire Railways Union Railway	1874
152	Northampton & Blisworth Railway	1874
153	London & North Western Railway. Additional Powers	1875
154	Midland Railway. New Works &c. Rushton and Bedford Widening Deviation	1875
155	Duston Minerals & Northampton & Gayton Junction Railway	1875
156	Braceborough Water	1875
157	Midland Railway. New Works. County of Northampton	1876
158	Great Northern Railway	1876
159	London & North Western Railway. New Works and Additional Lands	1876
160	Great Eastern Railway. Northern Extension	1877
161	Daventry & Weedon Railway	1877
162	Market Deeping Railway	1877

163	London & North Western Railway. Additional Powers	1877
164	Midland Railway. Additional Powers	1877
165	Midland Railway. Additional Powers	1878
166	Wellingborough Gas Works	1878
167	London & North Western Railway. Additional Powers	1878
168	Daventry & Weedon Railway	1878
169	Easton Neston Mineral & Towcester Roade & Olney Junction Railway	1878
170	London & North Western Railway	1879
171	Northampton Tramways	1879
172	Peterborough Tramways	1879
173	London & North Western Railway and Midland Railway. Market Harborough New Line and Works	1880
174	Midland Railway. Additional Powers	1880
175	London & North Western Railway. New Railways	1880
176	Daventry & Weedon Railway	1880
177	Kettering Gas	1881
178	Midland Railway. Additional Powers	1881
179	Northampton Tramways. Extensions	1881
180	Northampton Corporation. Commons, Streets, Roads Bill	1881
181	London & North Western Railway	1881
182	Midland Railway. Additional Powers	1882
183	Great Northern Railway	1883
184	London & North Western Railway	1883
185	Northampton & Daventry Railway	1883
186	Northampton Water	1883
187	Stratford-upon-Avon, Towcester & Midland Junction Railway. Deviation	1884
188	London & North Western Railway	1884
189	Northampton & Banbury & Metropolitan Junction Railway	1884
190	Northampton Daventry & Leamington Railway	1884
191	Kettering Water Works	1885
192	Wolverton & Stony Stratford Tramways. Deanshanger Extension	1886
193	Midland Railway. Additional Powers	1887
194	Kettering Waterworks	1887
195	Kettering Water	1888
196	Worcester & Broom Railway. Extension to Aylesbury	1888
197	London & North Western Railway	1888
198	Wellingborough & District Tramroads	1888
199	Towcester & Buckingham Railway	1888

200	Northampton Electric Lighting	1889
201	Midland Railway	1889
202	London & North Western Railway	1889
203	Metropolitan Railway. Extension to Moreton Pinkney	1889
204	Great Northern Railway. Various Powers	1889
205	Wellingborough & District Tramroads. Extensions	1889
206	Northampton Electric Light and Power Co. Ltd	1889
207	Midland Railway. New Lines &c.	1890
208	Great Northern Railway	1890
209	Manchester, Sheffield & Lincolnshire Railway. Extension to London Etc	1890
210	Crowland Railway	1890
211	London & North Western Railway. Additional Powers	1890
212	Midland Railway	1891
213	Manchester, Sheffield & Lincolnshire Railway. Extension to London Etc	1891
214	Northampton Tramways	1891
215	Midland Railway	1892
216	London & North Western Railway	1892
217	London & North Western Railway	1893
218	Manchester, Sheffield & Lincolnshire Railway. Deviation Railways Etc.	1893
219	Great Northern Railway	1894
220	Midland Railway	1894
221	Manchester, Sheffield & Lincolnshire Railway	1894
222	Kettering Electric Lighting	1895
223	Manchester, Sheffield & Lincolnshire Railway	1895
224	London & North Western Railway	1896
225	Great Northern Railway	1896
226	Manchester, Sheffield & Lincolnshire Railway	1896
227	Peterborough Electric Lighting	1897
228	Great Central Railway	1897
229	London & North Western Railway	1897
230	Midland Railway	1898
231	Nene Valley Waterworks	1898
232	Great Northern Railway	1898
233	Great Central Railway	1898
234	Buckingham, Towcester & Metropolitan Junction Railway	1899
235	London & North Western Railway	1899
236	Great Northern Railway	1899

237	Higham Ferrers Water Co.	1899
238	Irthlingborough Gas Co.	1899
239	Wellingborough & District Tramroads	1899
240	Wellingborough Electric Lighting	1899
241	Rothwell (Northampton) Gas Provisional Order	1899
242	Northampton Corporation Tramways	1900
243	Kettering Urban District Water	1900
244	Wellingborough Electric Lighting	1900
245	London & North Western Railway	1900
246	London & North Western Railway	1901
247	Midland Railway	1901
248	Finedon Urban District Water	1901
249	Higham Ferrers & Rushden Water Board	1901
250	Wellingborough Gas	1901
251	Great Northern Railway	1902
252	Midland Railway	1902
253	Wellingborough & District Tramroads	1902
254	Midland Railway	1903
255	Great Northern Railway	1903
256	Kettering Improvement. Widenings Grange Road	1903
257	Finedon Gas	1903
258	Northampton Electric Lighting	1903
259	Rushden, Higham Ferrers, Irthlingborough, and Thrapston (Rural) Electric Lighting	1904
260	Wellingborough & District Tramroads and Electricity Supply	1904
261	Great Western Railway (Additional Powers)	1904
262	Great Western Railway (New Railways)	1904
263	Midland Railway	1904
264	Higham Ferrers, Rushden and Wellingborough (Rural District) Electric Lighting	1905
265	Peterborough Gas	1905
266	Raunds Gas	1905
267	Wollaston Gas	1905
268	Great Northern Railway	1905
269	Great Northern Railway	1905
270	Great Western Railway (General Powers)	1908
271	Northampton Corporation Tramways Extension	1908
272	Stratford-upon-Avon & Midland Junction Railway	1909
273	Burton Latimer Water	1907

274	Northampton Corporation. Tramways and Street Widenings	1910
275	Great Northern Railway	1910
276	Midland Railway (Midland and Great Northern Railways Joint Committee)	1910
277	Rushden & District Electric Lighting	1911
278	Great Northern Railway	1912
279	Great Northern Railway	1912
280	Wellingborough Electric Lighting	1912
281	Northampton Corporation Water	1912
282	[Northampton Corporation Water]	[1919]
283	Northampton Corporation (Water)	1921
284	Midland Railway	1922
285	Rugby Urban District Council. General Powers Bill	1922
286	Northampton Electricity Extension	1922
287	Rushden & District Electricity Extension	[1923]
288	Wellingborough Electricity Extension	[1923]
289	Kettering Electricity Extension	1923
290	Northampton Electricity (Extension)	1924
291	Rushden & District Electricity (Extension)	1924
292	Northampton Gas	1925
293	Brackley Electricity	1927
294	Northampton Electricity (Extension)	[1928]
295	Mid-East England Electricity	1929
296	Kettering Electricity (Extension) Special Order, 1929	[1929]
297	Stamford Rural Electricity Special Order, 1930	1930
298	Northampton Gas Special Order, 1930	1930
299	Corby (Northants.) & District Water	1930
300	Grand Union Canal	1930
301	London & North Eastern Railway	1930
302	Grand Union Canal (Leicester Canals Purchase &c.)	1931
303	Bye-laws as to Petroleum Filling Stations made by the Northamptonshire County Council under Section 11 of the Petroleum (Consolidation) Act, 1928	1931
304	Kettering Gas Special Order, 1934	1934
305	Northampton Gas Special Order, 1933	[1932]
306	Northampton Gas, 1927	1927
307	Wellingborough Gas, 1935	[1935]
308	Banbury Gas, 1936	[1936]
309	Northampton Gas, 1937	1936
310	Stony Stratford Gas Special Order, 1937	1937

311	London & North Eastern Railway	1937
312	Wellingborough Gas Special Order, 1938	[1938]
313	United District Gas Order	[1938]
314	Kettering Gas Order	[1942]
315	Stony Stratford Gas Special Order, 1943	[1943]
316	Northampton Corporation Bill	1942
317	British Transport Commission	1954
318	Great Ouse Water	1960

NORTHAMPTONSHIRE MAIN SERIES PLANS

CATALOGUE

1. London & Birmingham Railway. Deposited 30 Nov. 1830. No Act.

From Paddington, through Chelsea, Willesden, Harrow, Kingsbury, Pinner (Middlesex); Watford, Oxhey, Abbots Langley, Rickmansworth, Hemel Hempstead, Great Gaddesden (Herts.); Edlesborough (Bucks.); Studham (Beds.), Little Gaddesden (Herts.); Eaton Bray, Leighton Buzzard (Beds.); Linslade (Bucks.); Great Brickhill, Stoke Hammond, Bletchley, Woughton on the Green, Loughton, Bradwell, Bradwell Abbey, Wolverton, Haversham, Castlethorpe, Hanslope (Bucks.); Hartwell, Ashton, Roade, Courteenhall, Milton, Blisworth, Gayton, Rothersthorpe, Kislingbury, Bugbrooke, Upper Heyford, Flore, Norton, Whilton, Long Buckby, Watford, Welton, Ashby St Ledgers, Kilsby (Northants.); Hillmorton, Rugby, Bilton, Newbold upon Avon, Long Lawford, Church Lawford, Wolston, Binley, Coventry, Stoneleigh, Allesley, Berkswell, Meriden, Hampton in Arden, Bickenhill, Elmdon, Solihull, Sheldon (Warwicks.); Yardley (Worcs.); Aston, Birmingham (Warwicks.).

Branch from near Blisworth to Northampton. Blisworth, Milton, Wootton, Hardingstone, Duston (Northants.).

2. London & Birmingham Railway. George Stephenson & Son, engineers. Deposited 30 Nov. 1831. No Act.

From St Pancras (Middlesex) to Birmingham (Warwicks.) almong similar route to **1** and **3**.

3. Proposed Rail-Way from London to Birmingham. George Stephenson & Son, engineers. Deposited 30 Nov. 1832. Act 3 & 4 Wm IV c. xxxvi (1833).

From St Pancras, through St Marylebone, Chelsea, Kensington, Fulham, Hammersmith, Acton, Willesden, Harrow, Pinner; Watford, Bushey, Abbots Langley, Kings Langley, Hemel Hempstead, Northchurch, Berkhampstead, Aldbury, Tring (Herts.); Marsworth, Pitstone, Cheddington, Ivinghoe, Mentmore, Grove, Linslade, Soulbury, Stoke Hammond, Bletchley, Woughton on the Green, Loughton, Bradwell, Bradwell Abbey, Wolverton, Haversham, Castlethorpe, Hanslope (Bucks.); Ashton, Hartwell, Courteenhall, Blisworth, Gayton, Bugbrooke, Nether Heyford, Stowe Nine Churches, Weedon Beck, Dodford, Norton, Brockhall, Whilton, Long Buckby, Watford, Ashby St Ledgers, Kilsby, Barby (Northants.); Hillmorton, Clifton on Dunsmore, Rugby, Bilton, Newbold upon Avon, Long Lawford, Church Lawford, Wolston, Willenhall, Binley, Coventry, Stoneleigh, Berkswell, Hampton in Arden, Barston, Bickenhill, Sheldon (Warwicks.); Yardley (Worcs.); Aston, Birmingham (Warwicks.).

4. London & Birmingham Railway. Robert Stephenson, engineer. Deposited 29 Nov. 1834. Act 5 & 6 Wm IV c. lvi (1835).

Proposed extension of the railway from the station in the Hampstead Road to Euston Grove in St Pancras (Middlesex).

Deviation from Stowe to Long Buckby (Northants.): in Stowe, Nether Heyford, Weedon Beck, Dodford, Brockhall, Norton, Whilton, Long Buckby, Watford.

Deviation from Watford (Northants.) to Hillmorton (Warwicks.): in Watford, Long Buckby, Ashby St Ledgers, Kilsby, Barby (Northants.); Hillmorton (Warwicks.).

Proposed diversion of the river Ouse in Wolverton and Haversham (Bucks.).

Proposed diversion of the river Avon in Wolston (Warwicks.).

5. London & Birmingham Railway. Deviation of line of railway from Watford (Northants.) to Hillmorton (Warwicks.). Robert Stephenson, engineer. Deposited 30 Nov. 1835.

In Watford, Ashby St Ledgers, Kilsby, Barby (Northants.); Hillmorton (Warwicks.).

6. Northern & Eastern Railway. Plan and section of the extended line from Cambridge to York. No engineer named. Deposited 30 Nov. 1836. No Act.

From London & Cambridge Railway at Trumpington, through Grantchester, Cambridge, Chesterton, Girton, Impington, Histon, Oakington, Long Stanton St Michael, Long Stanton All Saints, Over, Willingham (Cambs.); Bluntisham cum Earith, Colne, Somersham, Pidley cum Fenton, Warboys, Ramsey (Hunts.); Whittlesey St Mary & St Andrew (Cambs.), Stanground (Cambs. & Hunts.); Peterborough, Eye, Paston, Newborough (Northants.); Deeping St James, Langtoft, Baston, Thurlby, Bourne, Morton, Hacconby, Dunsby, Rippingale, Dowsby, Aslackby, Pointon, Sempringham, Billingborough, Horbling, Swaton, Helpringham, Burton, Little Hale, Great Hale, Heckington, Burton Pedwardine, Asgarby, Ewerby, Anwick, Ruskington, Dorrington, Digby, Rowston, Scopwick, Timberland, Martin, Blankney, Metheringham, Dunston, Nocton, Potter Hanworth, Branston, Heighington, Washingborough, Canwick, Boultham, Skellingthorpe (Lincs.); Thorney (Notts.); Broadholme, Saxilby cum Ingleby, Torksey, Brampton (Lincs.); North Leverton with Habblesthorpe, Littleborough, Sturton le Steeple, West Burton, Bole, Saundby, Beckingham, Walkeringham, Misterton (Notts.); Haxey (Lincs.); Misson (Notts.); Wroot (Lincs.); Finningley (Notts.); Cantley, Hatfield, Stainforth, Kirk Bramwith, Campsall, Fenwick, Fishlake, Sykehouse, Snaith & Cowick, Balne, Heck, Pollington, Kellington, Eggbrough, Hensall, Birkin, Chapel Haddlesey, Brayton, Gateforth, Hambleton, Wistow, Cawood, Ryther cum Ossendyke, Stillingfleet with Moreby, Acaster Selby, Bolton Percy, Appleton Roebuck, Acaster Malbis, Copmanthorpe, Askham Bryan (Yorks.); terminating in a junction with the York & North Midland Railway in York.

7. North & South Junction Railway. Francis Giles, engineer. Deposited 30 Nov. 1836. No Act.

From Oxford, through Wolvercote, Kidlington, Water Eaton, Thrup, Hampton Poyle, Hampton Gay, Blechingdon, Kirtlington, Tackley, Lower Heyford, Steeple Aston, Upper Heyford, Somerton, North Aston, Deddington, Souldern, Adderbury (Oxon.); Aynho, King's Sutton, Warkworth, Middleton Cheney, Chalcombe (Northants.); Cropredy, Bourton, Clattercote, Claydon (Oxon.); Farnborough, Wormleighton, Ladbroke, Southam, Stockton, Long Itchington, Birdingbury, Marton, Frankton, Princethorpe, Stretton on Dunsmore, Ryton on

Dunsmore, Wolston, Brandon & Bretford (Warwicks.).

8. South Midland Counties Railway, with branches to Leicester and Stamford. Francis Giles, engineer. Deposited 30 Nov. 1836. No Act.

Main Line: From London & Birmingham Railway at Courteenhall, through Milton, Collingtree, Wootton, Hardingstone, Duston, Northampton St Peter, Northampton All Saints, Northampton St Sepulchre, Northampton Extra-Parochial Lands (all town parishes on enlarged scale), Kingsthorpe, Boughton, Pitsford, Brixworth, Lamport, Draughton, Maidwell, Kelmarsh, Arthingworth, Great Oxendon, East Farndon, Little Bowden (Northants.); Great Bowden, Market Harborough, Lubenham, Foxton, Smeeton Westherby, Kibworth Beauchamp, Kibworth Harcourt, Burton Overy, Newton Harcourt; terminating in a junction with the Midland Counties Railway at Wigston Magna (Leics.).

Enlarged plans of Far Cotton, Kingsthorpe, Oxendon, Market Harborough, Kibworth, Newton Harcourt and elsewhere.

Branch from main line at Wigston Magna to Leicester & Swannington Railway at Leicester: Aylestone, Leicester St Margaret, Leicester St Mary, Borough of Leicester, The Newarke extra-parochial, Leicester Abbey Lands extra-parochial (Leics.); enlarged plan of Leicester town parishes.

Branch from main line at Lubenham to Stamford: Lubenham, Great Bowden, Weston by Welland, Medbourne (Leics.); Ashley (Northants.); Bringhurst, Drayton, Great Easton (Leics.); Caldecott, Liddington, Thorpe by Water, Seaton, Morcott, South Luffenham, North Luffenham, Ketton, Tinwell (Rutland); Stamford All Saints (Lincs.). Enlarged plans of Drayton, Great Easton, Caldecott, Morcott, Luffenham, Ketton, Tinwell.

9. South Midland Counties Railway, with branches to Leicester and Stamford. Francis Giles, civil engineer. Deposited 1 March 1837. Identical to **8**. No Act.

10. Railway commencing at a junction with the London & Birmingham Railway in Gayton; terminating near Peterborough. Robert Stephenson, engineer. Charles F. Cheffins, surveyor. Deposited 30 Nov. 1842. Plan sheets are marked 'N. & P.' (i.e. Northampton & Peterborough Railway). Act 6 & 7 Vict. c. lxiv (1843).

From Gayton, through Blisworth, Milton, Rothersthorpe, Hardingstone, Northampton St Giles, Great Houghton, Little Houghton, Brafield on the Green, Cogenhoe, Whiston, Castle Ashby, Earls Barton, Great Doddington, Wollaston, Irchester, Irthlingborough, Rushden, Higham Ferrers, Chelveston cum Caldecott, Stanwick, Raunds, Ringstead, Great Addington, Woodford near Thrapston, Denford, Islip, Thrapston, Titchmarsh, Thorpe Achurch, Lilford cum Wigsthorpe, Barnwell All Saints, Barnwell St Andrews, Oundle, Glapthorn, Cotterstock, Southwick, Fotheringhay, Nassington (Northants.); Elton, Sibson cum Stibbington (Hunts.); Castor, Ailsworth, Sutton (Northants.); Alwalton, Orton Waterville, Orton Longville, Woodstone, Fletton (Hunts.).

Alternative Line: Oundle, Ashton, Tansor, Warmington, Fotheringhay (Northants.)

10a. River Nene Improvements &c. Plans & sections of the River Nene, from Cross Keys Bridge to Woodstone Staunch, above Peterborough Bridge, together with the improvements proposed in the same, also of the intended dock in front of the town of Wisbech, the new cut from the Horse Shoe to Rummers Mill Sluice, the cut from the old channel at Barton Lane End to join the new cut, the alteration of the river at Guyhirn, the improvement in Moretons Leam for the drainage of Moretons Leam Wash, the improvement in Bevil's Leam, the new drain from Pond's Bridge to Caldicote Dyke & Monks Lode, the improvement of Caldicote Dyke & Monks Lode, together with the proposed new catchwater drain from the river Nene, near Standground Sluice, to Wells or Forty Foot Bridge, on Vermuydens Drain. Sir John Rennie, engineer. Deposited 30 Nov. 1840. No Act.

Plan of the Nene Navigation from Woodstone Staunch above Peterborough Bridge, to Cross Keys Bridge, shewing the proposed improvements. Includes Peterborough (Northants.); Woodstone, Fletton, Farcet (Hunts.); Stanground (Cambs. & Hunts.); Whittlesey St Mary & St Andrew, Wisbech, Leverington (Cambs.); Walsoken, West Walton (Norfolk); Tid St Giles (Cambs.); Tid St Mary, Sutton St Mary (Lincs.).

Plan of the line of the proposed catchwater drain from Standground Sluice on the river Nene to near Wells Bridge on the Forty Foot or Vermuydens Drain. Includes Farcet, Yaxley, Stilton, Caldecote, Denton, Holme, Connington, Sawtry All Saints, Sawtry St Andrews, Sawtry St Judith, Wood Walton, Great Raveley, Upwood, Bury, Ramsey (Hunts.).

Plan of the line of the proposed drain from the top of Caldicote Dyke along Caldicote Dyke, across Whittlesea Mere, and along Bevil's Leam, to the river Nene above Guyhirn. Includes Stilton, Caldecote, Denton, Yaxley, Holme, Ramsey, Farcet (Hunts.); Whittlesey St Mary & St Andrew, March, Wisbech (Cambs.).

Plan of the line of the proposed drain, from the top of Monk's Lode along Monk's Lode to Bevil's Leam, at Pond's Bridge. Includes Wood Walton, Sawtry St Andrew, Connington, Higney cum Ramsey, Holme cum Gratton, Ramsey (Hunts.); Whittlesey St Mary & St Andrew (Cambs.).

11. Oxford & Rugby Railway. 'J.K. Brunel' (i.e. Isambard Kingdom Brunel), engineer. Deposited 30 Nov. 1844. Act 8 & 9 Vict. c. clxxxviii (1845).

From junction with London & Birmingham Railway in Clifton on Dunsmore, through Rugby, Hillmorton, Bilton, Dunchurch, Leamington Hastings, Stockton, Napton on the Hill, Southam, Ladbroke, Chapel Ascote, Bishops Itchington, Fenny Compton, Farnborough (Warwicks.); Claydon, Clattercote, Cropredy, Bourton (Oxon.); Warkworth, Middleton Cheney, King's Sutton (Northants.); Adderbury (Oxon.); Aynho (Northants.); Souldern, Fritwell, Somerton, Upper Heyford, Steeple Aston, Lower Heyford, Rousham, Kirtlington, Tackley, Bletchington, Shipton on Cherwell, Hampton Gay, Kidlington, Begbroke, Yarnton, Wolvercote, Oxford (Oxon.); North Hinksey, South Hinksey (Berks.); ends in junction with Great Western Railway at Oxford.

12. London & York Railway. Alternative line [to that shown in **13** but numbered first] commencing at 76 miles 5 furlongs and terminating at 82 miles 5 furlongs from Old St Pancras Road at King's Cross, London. No engineer named. Deposited 30 Nov. 1844. No Act.

Starts in Peterborough; continues through Paston, Walton, Werrington, Marholm, Glinton, Etton, Helpston, ending at Lolham Bridges (Northants.).

17 is a duplicate of **12**.

13. Direct Northern Railway from London to York by way of Lincoln. Sir John Rennie, W. Gravatt, engineers. Surveyors: Sherrard & Hall, London (Grantham to Bawtry), W.F. Fairbank, Sheffield (Peterborough to Boston; Bawtry to Sheffield); Martin, Johnson & Fox, Leeds and London (Boston to Bawtry; Doncaster to Wakefield). Deposited 30 Nov. 1844.

No Act.

Part I. London to Grantham, with branches to Bedford and Stamford, and alternative line from Hitchin through Biggleswade to Sandy. Corresponds to sheets of plan in Book I, Nos 1–54. From St Pancras, through Islington, Hornsey, Tottenham, Edmonton, Friern Barnet (Middlesex); East Barnet (Herts.); Monken Hadley, Enfield, South Mimms (Middlesex); North Mimms, Hatfield, Digswell, Welwyn, Datchworth, Knebworth, Stevenage, Great Wymondley, Little Wymondley, Ippollitts, Hitchin, Walsworth, Ickleford (Herts.); Holwell, Shillington, Arlesey, Henlow, Clifton, Old Warden, Biggleswade, Northill, Sandy, Everton (Beds.); Tetworth (Hunts.); Tempsford, Little Barford (Beds.); Eynesbury, St Neots, Great Paxton, Offord D'Arcy, Buckden, Offord Cluny, Godmanchester, Brampton, Huntingdon, Great Stukeley, Abbots Ripton, Wood Walton, Sawtry St Judith, Sawtry All Saints & St Andrew, Connington, Glatton, Holme, Denton, Caldecote, Stilton, Yaxley, Farcet, Fletton, Woodstone (Hunts.); Peterborough, Walton, Paston, Marholm, Glinton, Peakirk, Etton, Helpston, Maxey, Ufford, Bainton (Northants.); Tallington, Uffington, Barholm, Greatford, Braceborough (Lincs.); Essendine (Rutland); Carlby, Careby, Little Bytham, Creeton, Swayfield, Swinstead, Corby, Burton Coggles, Bitchfield, Basingthorpe, Boothby Pagnell, Great Ponton, Little Ponton, Grantham, Somerby, Harrowby (Lincs.).

Alternative Line from Hitchin through Biggleswade to Sandy. From Hitchin, through Walsworth (Herts.); Holwell (Beds.); Ickleford (Herts.); Arlesey, Henlow, Langford, Biggleswade, Northill, to Sandy (Beds.).

Branch to Bedford. From Sandy, through Blunham, Moggerhanger, Willington, Cople, Cardington, Eastcotts, Goldington, Bedford (Beds.).

Branch to Stamford. From Maxey, through Ufford, Bainton (Northants.); Tallington, Uffington (Lincs.); Barnack, Pilsgate, Stamford Baron (Northants.); Stamford (Lincs.).

Part II. Grantham to Bawtry. Plan sheets 54–74 in Book I. From Grantham, through Manthorpe, Great Gonerby, Belton, Syston, Barkston, Marston, Hougham, Westborough, Dry Doddington, Stubton, Claypole (Lincs.); Balderton, Newark on Trent, South Muskham, North Muskham, Cromwell, Norwell, Carlton on Trent, Sutton upon Trent, Grassthorpe, Marnham, Normanton on Trent, Fledborough, Darlton, East Markham, East Drayton, Askham, Headon cum Upton, Eaton, Ordsall, West Retford, Babworth, Sutton cum Lound, Blyth, Barnby Moor, Torworth, Ranskill, Scrooby, Scaftworth, Everton (Notts.); Bawtry (Yorks.).

Part III. Bawtry to York. Plan Part III sheets 75–94a. From Bawtry, through Austerfield (Yorks.); Harworth (Notts.); Rossington, Loversall, Cantley, Warmsworth, Doncaster, Wheatley, Bentley with Arksey, Langthwaite with Tilts, Thorpe in Balne, Barnby Dun, Owston, Burghwallis, Moss, Campsall, Fenwick, Balne, Snaith & Cowick, Pollington, Heck, Hensall, Temple Hirst, Birkin, Burn, Brayton, Selby, Wistow, Cawood, Acaster Selby, Acaster Malbis, Bishopthorpe, Middlethorpe, Dringhouses, Yorks (Yorks.).

Side Branch at Doncaster. In Bentley with Arksey and Doncaster (Yorks.).

Part IV. Lincoln Line, from Peterborough via Spalding, Boston, Lincoln and Gainsborough to Bawtry. Sheets of Plan IV numbered Lincoln 1 to Lincoln 45. From Marholm, through Werrington, Peakirk, Glinton, Maxey (Northants.); Deeping St James, Deeping Fen, Crowland, Spalding, Pinchbeck, Surfleet, Gosberton, Sutterton, Algarkirk, Kirton, Frampton, Wyberton, Skirbeck, Boston, Frithville, Sibsey, Langriville, Coningsby, Fishtoft, Brothertoft, Kirkstead, Woodhall, Fosdyke, Swineshead, Dogdyke, Billinghay, Tattershall, Tattershall Thorpe, Timberland, Martin, Thornton le Fen, Woodhall, Thimbleby, Edlington, Stixwould, Linwood, Blankney, Metheringham, Horsington, Bucknall, Dunston, Bardney, Nocton, Potter Hanworth, Branston, Washingborough, Fiskerton, Heighington, Cherry Willingham, Greetwell, Canwick, Monk's Liberty, Lincoln, Boultham, Skellingthorpe, Burton by Lincoln, Saxilby cum Ingleby, Torksey, Hardwick, Stowe, Marton, Gate Burton, Knaith, Lea, Gainsborough, Beckingham (Lincs.); Saundby, Walkeringham, Gringley on the Hill, Everton, Scaftworth (Notts.).

Side Branch to the Shipping Quay at Boston. In Skirbeck and Boston (Lincs.).

Part V. Sheffield Line (Bawtry to Sheffield); Wakefield Line (Doncaster to Wakefield). Plan Part V sheets Sheffield 1–12; Wakefield 1–9. From Bawtry (Yorks.); through Harworth (Notts.); Tickhill, Maltby, Hooton Levitt, Laughton en le Morthen, Stainton with Hellaby, Treeton, Brampton en le Morthen, Wickersley, Bramley, Braithwell, Brinsworth, Rotherham, Whiston, Tinsley, Attercliffe cum Darnall, Brightside Bierlow, Sheffield; [Wakefield Line:?] Bentley with Arksey, Langthwaite with Tilts, Adwick le Street, Hampole, Hamphall Stubbs, Hooton Pagnell, South Elmsall, South Kirkby, North Elmsall, Hemsworth, Wintersett, Wragby, Ryhill, Huntwick woth Foulby & Nostell, Walton, Sandal Magna Crofton (Yorks.).

Side Branch. In Crofton and Warmfield cum Heath (Yorks.).

Main Branch. In Warmfield cum Heath, Sandal Magna, Wakefield (Yorks.)

Plans deposited in Northants. cover only the main line from London to York and the branch to Bedford.

14. London & Worcester & Rugby & Oxford Railway with an extension from Worcester to Dudley [and Wolverhampton crossed through**].** No engineer named. Deposited 30 Nov. 1844. No Act.

From junction with the London & Birmingham Railway in Marsworth, through Cheddington, Ivinghoe (Bucks.), Tring (Herts.), Hulcott, Bierton with Broughton, Aylesbury, Quarrendon, Fleet Marston, Waddesdon, Woodham, Quainton, Grendon Underwood, Edgcott, Marsh Gibbon (Bucks.); Ambrosden, Blackthorn, Launton, Bicester, Bicester Market End, Bicester King's End, Chesterton, Bucknell, Middleton Stoney, Ardley, Upper Heyford, Somerton, Fritwell, Souldern (Oxon.); Aynho, King's Sutton (Northants.); Adderbury, Boddicote (Oxon.); Marston St Lawrence, Warkworth (Northants.); Banbury, Neithrop (Oxon.); Middleton Cheney, Chalcombe (Northants.); Cropedy, Wardington, Bourton, Cropedy, Claydon, Clattercote (Oxon.); Farnborough, Fenny Compton, Burton Dassett, Kineton, Butlers Marston, Pillerton Hersey, Pillerton Priors, Eatington (Warwicks.), Alderminster (Worcs.); Preston on Stour (Gloucs.); Atherstone on Stour (Warwicks.); Clifford Chambers (Gloucs.); Weston upon Avon (Warwicks.), Welford on Avon (Warwicks.); Long Marston (Gloucs.); Dorsington (Warwicks.); Pebworth (Worcs.); North & Middle Littleton, South Littleton, Bretforton, Offenham, Badsey, Aldington, Bengeworth, Evesham All Saints, Evesham St Lawrence, Great & Little Hampton, Cropthorne, Fladbury, Pershore, Pinvin, Stoulton, Norton juxta Kempsey, Worcester, North Claines, Ombersley, Hartlebury, Kidderminster, Lower Mitton, Upper Mitton, Wolverley, (Worcs.); Kinver (Staffs.); Wollaston (Worcs.); Amblecote, Kingswinford (Staffs.); Dudley (Worcs.).

Reference to the line from Bucknell (Oxon.) to St Aldate near Oxford (Berks.): Bucknell, Bicester, Chesterton, Weston on the Green, Oddington, Islip, Kidlington, Water Eaton,

Cutteslowe, Oxford (Oxon.); North Hinksey (Berks.), Oxford St Aldate (Grand Pont liberty) (Oxon. and Berks.).

Reference to the line from Fenny Compton (Warwicks.) to the London & Birmingham Railway at Hillmorton: Fenny Compton, Wormleighton, Watergall, Upper Hodnell, Ladbroke, Southam, Napton on the Hill, Grandborough, Caldecote (Warwicks.); Barby, Onley (Northants.); Hillmorton (Warwicks.).

Plans and sections lack cover sheets giving name of engineer. First series of sheets (marked Oxford or Ox. & R.) shows lines starting from L&B in Marsworth (Bucks.) and ending at Barneshall in parish of St Peter, Worcester. Second plan (with new deposit memorial endorsed; no abbreviated running title) is a continuation through Worcester to Dudley, terminating in junction with proposed Dudley to Wolverhampton railway. Third set (marked Oxford [Branch]) shows branch from junction with main line at Bucknell to junction with GWR at Oxford. Fourth set (marked Rugby Branch) shows branch from Fenny Compton to L&B at Hillmorton. Followed by sections for all four sets of plans.

The label on the wrapper for this map (partly torn away) reads: ' ... Nov. 1844. L&NW Ry Co. London & Worcester and Rugby & Oxford Ry Co.'.

15. Midland Railways (sic) Extensions. Syston & Peterborough Railway. George Stephenson, Charles Liddell, engineers. J.G. Binns, surveyor. Deposited 30 Nov. 1844. Act 8 & 9 Vict. c. lvi (1845).

From junction with Midland Railway near Syston station, through Syston, Queniborough, Rearsby, Thrussington, Hoby, Brooksby, Rotherby, Frisby on the Wreak, Kirby Bellars, Asfordby, Melton Mowbray, Thorpe Arnold, Brentingby & Wyfordby, Stapleford, Saxby, Wymondham, Edmondthorpe (Leics.); Whissendine, Teigh, Ashwell, Oakham, Burley, Manton, Wing, Pilton, North Luffenham, South Luffenham, Ketton (Rutland); Easton on the Hill (Northants.); Tinwell (Rutland); Stamford (Lincs.); Stamford Baron (Northants); Uffington (Lincs.); Barnack, Ufford, Helpston, Etton, Glinton, Paston, Marholm, Peterborough (Northants.); Fletton (Hunts.); ending in junction with Eastern Counties Railway at Peterborough.

16. Cambridge & Lincoln Railway. G.W. Buck, engineer. J.W. Bazalgette, acting engineer. Deposited 30 Nov. 1844. No Act.

From a junction with the Eastern Counties Railway Extension to Ely, through Cambridge, Chesterton, Girton, Impington, Histon, Oakington, Long Stanton St Michael, Swavesey, Fen Drayton (Cambs.); Fen Stanton, Holywell cum Needingworth, St Ives, Woodhurst, Old Hurst, Wyton, Houghton, Broughton, Warboys, Wistow, Upwood, Bury, Ramsey, Farcet (Hunts.); Stanground (Cambs. & Hunts.); Peterborough, Paston, Werrington, Peakirk, Glinton, Northborough, Maxey (Northants.); Market Deeping, Langtoft, Baston, Thurlby, Bourne, Morton, Hacconby, Dunsby, Rippingale, Dowsby, Aslackby, Sempringham, Pointon, Billingborough, Horbling, Swaton, Helpringham, Great Hale, Little Hale, Heckington, Asgarby, Ewerby, Anwick, Ruskington, Dorrington, Digby, Rowston, Billingham, Walcot, Timberland, Martin, Scopwick, Blankney, Linwood, Metheringham, Dunston, Nocton, Potter Hanworth, Branston, Heighington, Washingborough, Canwick (Lincs.); ending in a terminus at Lincoln.

17. London & York Railway. Alternative line commencing at 76 miles 5 furlongs and terminating at 82 miles 5 furlongs from Old St Pancras Road, King's Cross. No engineer named. Deposited 30 Nov. 1844. No Act.

From a deviation from the original line in Peterborough, through Paston, Walton, Werrington, Marholm, Glinton, Etton, Helpston; rejoining main line at Lolham Bridges (Northants.). Duplicate of **12**.

18. London, Birmingham & Buckinghamshire Railway. No engineer named. Deposited 12.03 a.m. 1 Dec. 1845. No Act.

Plan and section of a line of railway from Uxbridge in Denham (Bucks.) to or near Banbury in Warkworth (Northants.). From a river crossing in Denham, through Iver, Chalfont St Peter, Chalfont St Giles, Beaconsfield, Coleshill, Little Missenden, Great Missenden, Wendover, Stoke Mandeville, Aylesbury, Quarrendon, Fleet Marston, Waddesdon, Quainton, Twyford, Preston Bisset, Chetwode, Barton Hartshorn (Bucks.); Shelswell, Newton Purcell, Mixbury (Oxon.); Evenley, Croughton, King's Sutton, Newbottle, Warkworth (Northants.).

Plan and sections made up from hand-drawn tracings pasted on to cartridge paper. No county rubrics. Pencilled note on back of last sheet: 'Mr Bramah, 6 Gt Winchester St. 3 mins. after 12'.

19. Boston, Stamford & Birmingham Railway, from Birmingham to Boston and Wisbech, with a branch to Market Harborough. Charles Vignoles, engineer. Deposited 30 Nov. 1845. Act 9 & 10 Vict. c. xciii (1846).

From a junction with the Midland Railway near Wigston Station, through Wigston Magna, Newton Harcourt, Burton Overy, Kibworth Beauchamp, Church Langton (Leics.); Weston by Welland, Ashley (Northants.); Bringhurst (Leics.); Cottingham, East Carlton (Northants.); Caldecott, Liddington, Seaton, Thorpe by Water, Morcott, South Luffenham (Rutland); terminating there in a junction with the Syston & Peterborough Railway, which is itself then planned through South Luffenham, Ketton, Tinwell (Rutland); Easton on the Hill (Northants.); Stamford (Lincs.); Stamford Baron, Barnack (Northants.); Uffington (Lincs.); Ufford, Helpston (Northants.); deviating there from the Syston & Peterborough Railway; continuing through Etton, Glinton, Peakirk, Newborough, Eye (Northants.); Thorney (Cambs.); Long Sutton, Sutton St Edmund (Lincs.); Leverington (Cambs.), Wisbech (Cambs.); terminating in a junction with the Wisbech branch of the Lynn & Ely Railway.

Branch to Market Harborough. From a terminus in Market Harborough, through Great Bowden, Church Langton (Leics.); terminating in a junction with the main line.

Boston Line. From the Wisbech line in Peakirk, through Maxey (Northants.); Deeping St James, Deeping Fen, Crowland, Spalding, Pinchbeck, Surfleet, Gosberton, Sutterton, Algarkirk, Kirton, Frampton, Wyberton, Boston Skirbeck Quarter (Lincs.); ending in a terminus at Boston.

20. Leicester & Bedford Railway. No engineer named. Deposited 30 Nov. 1845. No Act.

From a junction with the London & York Railway in Hitchin, through Ickleford, Pirton (Herts.); Shillington (Beds.); Holwell (Herts.); Upper Stondon, Meppershall, Campton, Shefford, Shefford Hardwick, Chicksands Priory, Southill, Haynes, Wilshamstead, Elstow, Bedford, Biddenham, Bromham, Clapham, Oakley, Milton Ernest, Pavenham, Felmersham, Sharnbrook, Souldrop, Wymington, Podington (Beds.); Irchester, Wellingborough, Irthlingborough, Finedon, Great Harrowden, Little Harrowden, Isham, Barton Seagrave, Pytchley, Kettering, Broughton, Cransley, Thorpe Malsor, Rothwell, Desborough, Braybrooke, Little Bowden (Northants.); Great Bowden, East Langton, West Langton, Foxton, Kibworth

Beauchamp, Kibworth Harcourt, Newton Harcourt, Burton Overy, Wigston Magna, Leicester (Leics.); terminating in a junction with the Midland Railway at Leicester.

With branches to join the Northampton & Peterborough Railway in Irthlingborough and Irchester (Northants.).

21. London & Birmingham Railway. Banbury Line. Deposited 30 Nov. 1845. No cover sheet; no engineer named. No Act.

From a junction with the main line in Gayton, through Tiffield, Towcester, Abthorpe, Bradden, Slapton, Wappenham, Weedon Lois, Helmdon, Stuchbury, Greatworth, Farthinghoe, Marston St Lawrence, Thenford, Middleton Cheney, King's Sutton, Newbottle, Warkworth (Northants.); ending at Banbury (Oxon.).

22. Warwickshire & London Railway. Plan and Section from Worcester to Weedon. Robert Stephenson, engineer. Deposited 30 Nov. 1845. No Act.

Main Line: From terminus in Worcester through North Claines, Warndon, Tibberton, Crowle, Oddingley, Huddington, Himbleton, Grafton Flyford, Kington, Dormston, Inkberrow, Rous Lench, Abbots Morton (Worcs.); Salford Priors, Bidford on Avon, Temple Grafton, Binton, Welford on Avon, Old Stratford & Drayton, Weston on Avon (Warwicks.); Clifford Chambers (Gloucs.); Alveston, Loxley (Warwicks.); Alderminster (Worcs.); Eatington, Pillerton Priors, Pillerton Hersey, Butlers Marston, Kineton, Burton Dassett, Fenny Compton, Wormleighton (Warwicks.); Cropredy (Oxon.); Boddington, Aston le Walls, Byfield, Woodford cum Membris, Charwelton, Preston Capes, Fawsley, Everdon, to Weedon Beck (Northants.), ending in junction with London & Birmingham Railway.

Alcester Branch: Bidford, Wixford, Arrow, Alcester (Warwicks.).

Droitwich Branch: Huddington, Oddingley, Himbleton, Droitwich St Peter, Dodderhill, Droitwich St Andrew, Droitwich St Nicholas (Worcs.).

Book of reference includes main line and both branches; plans and sections include only main line and Droitwich Branch.

23. Direct London & Manchester Railway, with a branch to Crewe. Sir John Rennie, John Urpeth Rastrick, George Remington, engineers. Deposited 29 Nov. 1845. No Act.

[*First Book*]: From Islington (Middlesex), through Hornsey, Tottenham, Edmonton, Friern Barnet (Middlesex); East Barnet, Chipping Barnet (Herts.); Monken Hadley, South Mimms (Middlesex); Ridge, St Albans, Shenley, Saundridge, Wheathampstead, Harpenden (Herts.); Luton, Stopsley, Limbury, Streatley, Barton in the Clay, Higham Gobion, Pulloxhill, Flitton, Maulden, Ampthill, Houghton Conquest, Wilshamstead, Elstow, Kempston (Beds.).

[*Second Book*]: From Heaton Norris (Lancs.), through Burnage, Didsbury, Withington, Hulme, Chorlton upon Medlock, Manchester (Lancs.).

[*Third Book*]: From Bedford (Beds.), through Biddenham, Clapham, Bromham, Oakley, Pavenham, Milton Ernest, Felmersham, Sharnbrook, Souldrop, Podington, Wymington (Beds.); Irchester, Irthlingborough, Wellingborough, Great Harrowden, Little Harrowden, Burton Latimer, Isham, Barton Seagrave, Pytchley, Kettering, Broughton, Cransley, Thorpe Malsor, Rothwell, Desborough, Braybrooke, Little Bowden (Northants.); Great Bowden, Foxton, East Langton, Kibworth Beauchamp, Smeeton Westerby, Burton Overy, Glen Magna, Newton Harcourt, Wistow, Wigston Magna, Knighton, Leicester, Glenfield, Braunstone, Kirby Muxloe, Braunston Frith, Kirby Frith, Ratby, Thornton, Stanton under Bardon, Ibstock, Dorrington (Leics.).

[*Fourth Book*]: *Crewe Branch.* From Checkley, Upper Tean (Staffs.), through Leigh, Huntley, Cheadle, Caverswall, Caverswall, Forsbrook, Dilhorne, Stoke upon Trent, Stoke upon Trent, Burslem, Wolstanton, Audley, Talke o'th'Hill (Staffs.); Alsager, Barthomley, Church Lawton, Monks Coppenhall (Cheshire).

[*Fifth Book*]: From Packington, through Ravenstone with Snibston, Normanton on the Heath, Nailstone, Ashby de la Zouch, Blackfordby, Ashby Woulds (Leics.); Church Gresley, Castle Gresley, Stanton & Newhall, Stapenhill (Derbys.); Burton upon Trent, Rolleston, Tutbury, Hanbury, Draycott in the Clay, Marchington, Doveridge, Uttoxeter, Checkley, Upper Tean, Leigh, Huntley, Cheadle, Kingsley, Consall, Cheddleton, Basford, Ipstones, Longsdon, Lowe, Leek, Leekfrith, Rudyard, Rushton Spencer (Staffs.); Bosley, Sutton, Macclesfield, Upton, Prestbury, Butley, Woodford, Gawsworth, Handforth, Bulkeley (Cheshire).

24. Direct Northern Railway from London to York, with branches. John Miller, engineer. Deposited 29 Nov. 1845. Act 9 & 10 Vict. c. lxxi. (1846).

Main Line: From terminus in St Pancras (Middlesex), through Islington, Hornsey, Tottenham, Friern Barnet (Middlesex); East Barnet (Herts.); Monken Hadley, Enfield, South Mimms (Middlesex); North Mimms, Hatfield, Digswell, Welwyn, Datchworth, Knebworth, Stevenage, Graveley, Little Wymondley, Great Wymondley, Letchworth (Herts.); Stotfold, Arlesey, Langford, Biggleswade, Sandy, Everton, Tempsford, Little Barford, Eaton Socon (Beds.); Little Paxton, Southoe, Diddington, Buckden, Brampton, Alconbury, Upton, Coppingford, Sawtry St Judith, Sawtry All Saints & St Andrews, Conington, Glatton, Denton, Caldecote, Stilton, Folksworth, Morborne, Haddon, Chesterton, Water Newton (Hunts.); Castor, Barnack (Northants.); Uffington (Lincs.); Ryhall, Essendine (Rutland); Carlby, Careby, Little Bytham, Creeton, Castle Bytham, Swayfield, Corby, Burton Coggles, Bassingthorpe, South Stoke, Great Ponton, Little Ponton, Grantham, Somerby, Great Gonerby, Belton, Syston, Barkston, Marston, Hougham, Hough on the Hill, Stubton, Fenton, Beckingham, Stapleford, Norton Disney, Swinderby (Lincs.); South Scarle (Notts.); North Scarle (Lincs.); North Clifton (Notts.); Newton, Kettlethorpe (Lincs.); Laneham, Rampton, Treswell, South Leverton, North Leverton with Habblesthorpe, Sturton le Steeple, West Burton, Bole, Saundby, Beckingham, Walkeringham, Misterton (Notts.); Owston Ferry, Haxey, Wroot (Lincs.); Hatfield, Thorne, Fishlake, Snaith & Cowick, Drax, Brayton, Selby, Wistow, Cawood, Stillingfleet with Moreby, Acaster Malbis, Bishopthorpe, York (Yorks.); ending at a junction with the Great North of England Railway.

Bedford Branch. From the main line in Stevenage (Herts.), through Great Wymondley, Little Wymondley, Ippollitts, Hitchin, Holywell, Ickleford (Herts.); Henlow, Clifton, Southill, Haynes, Cardington, Bedford (Beds.); ending at a terminus in Bedford.

Huntingdon Branch. From main line in Buckden (Hunts.), through Brampton, Huntingdon (Hunts.); terminating in a junction with the proposed Ely & Huntingdon Railway.

Peterborough Branch. From main line in Folksworth (Hunts.), through Morborne, Yaxley, Orton Longueville, Woodstone, Fletton (Hunts.); Barnack (Northants.); Uffington (Lincs.); terminating in a junction with the Midland Railway Syston & Peterborough branch (thus book of reference; plan names line as Northampton & Peterborough Railway).

Newark Branch. From main line in Fenton, through Beckingham, Claypole (Lincs.); Barnby in the Willows, Coddington, Balderton, Newark upon Trent, Kelham (Notts.); terminating in what appears to be a (projected?) Midland Railway branch.

Lincoln Branch. From a terminus in Lincoln, through Boultham, Skellingthorpe, Saxilby cum Ingleby (Lincs.); Thorney (Notts.); Torksey, Kettlethorpe (Lincs.); Laneham (Notts.); terminating in a junction with the main line.

Sheffield Branch. From main line in Rampton (Notts.), through Treswell, South Leverton, North Leverton with Habblesthorpe, Clarborough (Notts.); terminating in a junction with the proposed Sheffield & Lincolnshire Railway.

Leeds Branch. From main line in Misterton (Notts.), through Haxey (Lincs.); Gringley on the Hill, Misson (Notts.); Finningley (Notts.); Cantley, Doncaster, Bentley with Arksey, Adwick le Street, Owston, Burghwallis, Campsall, Kirk Smeaton, Badsworth, Darrington, Pontefract, Castleford, Featherstone, Methley, Ferry Frystone (Yorks.); terminating in junctions with the Midland Railway and the York & North Midland Railway.

Also a branch or side railway to a junction with the Leeds & Selby Railway through the parishes of Brayton and Selby (Yorks.) (included in the book of reference but no plan or section).

25. Northampton, Bedford & Cambridge Railway. William Purdon, engineer. Deposited 29 Nov. 1845. No Act.

From a terminus in Bridge Street, Northampton (enlarged plan), through Northampton (All Saints, St Giles), Abington, Great Houghton, Hardingstone (junction with Northampton & Peterborough Railway), Little Houghton, Brayfield on the Green, Denton, Yardley Hastings (Northants.); Olney Park Farm, Olney, Clifton Reynes, Newton Blossomville (Bucks.); Turvey, Stagsden, Kempston, Biddenham, Bedford (Beds.); terminating in a junction with the London & Birmingham and Bedford Railway.

26. Sheffield, Nottingham & London Direct Railway. Line of Railway from Nottingham to Chesterfield. No engineer named. Deposited 29 Nov. 1845. No Act.

From a junction with the Midland Railway in Nottingham St Mary, through Lenton, Radford, Basford, Bulwell, Papplewick, Hucknall Torkard, Linby, Newstead, Annesley, Kirkby in Ashfield (Notts.); Pinxton, South Normanton, Blackwell, Tibshelf, Morton, North Wingfield, Wingerworth, Chesterfield (Derbys.); terminating in a junction with the Midland Railway.

No part of the route passes through Northants.

27. London & York (Great Northern) Railway. Stamford & Spalding Branch. No engineer named. Deposited 30 Nov. 1845. Act 9 & 10 Vict. c. cccliii (1846).

From a junction with the Old Stamford branch of the London & York line in Bainton and crossing the London & York main line in Ufford (Northants.); through Tallington (Lincs.); Maxey (Northants.); Market Deeping, Deeping St James, Deeping Fen, Crowland (Lincs.); terminating in a junction with the London & York Railway.

28. Stamford, Market Harborough & Rugby Railway, with a branch to the Leicester & Bedford Railway. No engineer named. Deposited 30 Nov. 1845. No Act.

From a junction with the London & York Railway in Stamford (Lincs.), through Tinwell (Rutland); Easton on the Hill (Northants.); Ketton (Rutland); Collyweston, Duddington (Northants.); Tixover (Rutland); Wakerley (Northants.); Barrowden, Seaton (Rutland); Harringworth (Northants.); Thorpe by Water, Liddington (Rutland); Gretton (Northants.); Caldecott (Rutland); Great Easton, Bringhurst (Leics.); Cottingham, East Carlton, Ashley (Northants.); Medbourne, Slawston (Leics.); Weston by Welland (Northants.); Thorpe Langton, Great Bowden (Leics.); Little Bowden, Great Oxendon, Braybrooke, Clipston (Northants.); terminating in a junction a little short of Market Harborough (Leics.).

29. Midland & Eastern Counties Railway. John Miller, engineer. Deposited 29 Nov. 1845. No Act.

From a junction with the Northern & Eastern Railway, or Cambridge Line of the Eastern Counties Railway, in Cambridge, through Trumpington, Grantchester, Barton, Haslingfield, Harlton, Little Eversden, Comberton, Great Eversden, Toft, Kingston, Caldecote, Bourn, Caxton (Cambs.); Great Gransden (Hunts.); Croxton (Cambs.); Abbotsley, Eynesbury, St Neots (Hunts.); Eaton Socon, Colmworth, Bolnhurst, Thurleigh, Keysoe, Riseley, Bletsoe, Sharnbrook, Knotting, Souldrop, Podington (Beds.); Wollaston, Bozeat, Strixton, Grendon, Castle Ashby, Whiston, Cogenhoe, Ecton, Great Billing, Little Billing, Brafield on the Green, Weston Favell, Abington, Northampton (St Giles, All Saints) (enlarged plan of Bridge Street), Hardingstone, Duston, Upton, Kislingbury, Bugbrooke, Nether Heyford, Stowe Nine Churches, Weedon Beck, Everdon, Newnham, Badby, Catesby, Staverton, Hellidon (Northants.); Priors Marston, Napton on the Hill, Upper Radbourne, Ladbroke, Chapel Ascote, Bishops Itchington, Harbury, Chesterton & Kingston, Moreton Morrell, Newbold Pacey, Wellesbourne Hastings, Wellesbourne Mountford, Charlecote, Alveston, Old Stratford & Drayton (Warwicks.); Clifford Chambers (Gloucs.); Weston on Avon, Milcote, Welford on Avon, Long Marston, Dorsington (Warwicks.); Pebworth, North & Middle Littleton, South Littleton, Bretforton (Worcs.); terminating in a junction with the Oxford, Worcester & Wolverhampton Railway.

30. South Midlands & Southampton Junction Railway, from Northampton to Reading and Basingstoke. Charles E. Bernard, engineer. Deposited 30 Nov. 1845. No Act.

From a terminus in Basingstoke, through Eastrop, Monk Sherborne, Basing, Sherfield upon Loddon, Bramley, Silchester (Hants.); Stratfield Mortimer (Berks.); Mortimer West End (Hants.); Sulhampstead Banister, Beech Hill, Grazeley, Shinfield, Burghfield, Reading (Berks.); junction with Newbury branch of Great Western Railway; crosses GWR main line near Reading station; branches from GWR main line through Reading (Berks.); Caversham (Oxon.); Sonning (Berks.); Shiplake, Harpsden, Rotherfield Peppard, Henley on Thames, Bix (Oxon.); Fawley (Bucks.); Watlington, Pishill (Oxon.); Turville (Bucks.); Pyrton, Shirburn, Lewknor, Adwell, Attington, Thame (Oxon.); Haddenham, Long Crendon, Ashendon, Lower Winchendon, Upper Winchendon, Westcott, Wotton Underwood, Waddesdon, Quainton, Grendon Underwood, Middle Claydon, East Claydon, Steeple Claydon, Addington, Padbury, Adstock, Buckingham, Thornborough, Foscott, Leckhampstead (Bucks.); Wicken, Passenham, Cosgrove, Potterspury, Alderton, Paulerspury, Stoke Bruerne, Shutlanger, Tiffield, Blisworth (Northants.); terminating in a junction with the London & Birmingham Railway at Blisworth Station.

31. Northampton, Lincoln & Hull Direct Railway. Engineer J.U. Rastrick. Deposited 30 Nov. 1845. No Act.

[Part I.] From a junction with the Northampton and Peterborough branch at Northampton (Bridge Street) Station through Hardingstone, Duston, Northampton All Saints, Northampton St Peter, Northampton St Sepulchre, Dallington,

Kingsthorpe, Abington, Weston Favell, Moulton, Overstone, Sywell, Hannington, Orlingbury, Walgrave, Broughton, Pytchley, Cransley, Kettering, Weekley, Geddington, Newton, Great Oakley, Middleton hamlet, Cottingham, Rockingham (Northants.); Great Easton (Leics.); Caldecott, Liddington, Thorpe by Water, Seaton, Bisbrooke, Wing, Manton, Oakham, Gunthorpe, Hambleton, Egleton, Oakham, Burley, Cottesmore, Barrow, Market Overton, Thisleton (Rutland); South Witham, North Witham, Colsterworth, Easton, South Stoke, Great Ponton, Little Ponton, Somerby, Grantham (enlarged town plan), Harrowby, Manthorpe, Great Gonerby, Belton, Syston, Marston, Barkston, Hougham, Caythorpe, Fulbeck, Leadenham, Welbourn, Skinnand, Navenby, Boothby Graffoe, Coleby, Harmston, Waddington, Bracebridge, South Common, Lincoln (Lincs.); terminating at High Street in junctions with the Nottingham & Lincoln Railway, the proposed London & York Railway and proposed line from Market Rasen (i.e. Part II below).

Part II. From a junction with an unidentified railway in Middle Rasen, through Walesby, Market Rasen, Linwood, Buslingthorpe, Lissington, Wickenby, Snelland, Stainton by Langworth, Scothern, Sudbrooke, Barlings, Reepham, Fiskerton, Cherry Willingham, Greetwell, Monk's Liberty, Lincoln (Lins.); terminating in junctions at High Street as in [Part I] above.

32. Oxford, Witney, Cheltenham & Gloucester Independent Extension Railway, with two branches therefrom. B. Albano, R. Nicholson, engineers. Deposited 30 Nov. 1845. No Act.

Oxford Branch. From a terminus about three furlongs west of the Great Western main line in Oxford St Aldate (Grand Pont liberty) (Berks.), passing over the GWR, continuing through North Hinksey (Berks.) to Cowley (Oxon.), terminating in an end-on branch with main line as below.

Main Line. From an end-on junction with the Oxford Branch as above, through Cowley, Oxford St Clements, Iffley, Headington, Forest Hill with Shotover, Holton, Cuddesdon, Wheatley, Great Milton, Chilworth, Waterstock (Oxon.); Ickford (Bucks.); Tiddington, Albury, Thame, Towersey, Chinnor (Oxon.); Bledlow, Saunderton, Horsendon, Bradenham, West Wycombe, Chepping Wycombe, Wooburn, Beaconsfield, Burnham, Farnham Royal, Hedgerley Dean, Hedgerley, Fulmer, Langley Marish, Iver, Denham (Bucks.); Hillingdon, Uxbridge, Cowley, Hayes, Norwood (Middlesex); terminating in a junction with the Great Western Railway.

Bicester Branch. From main line in Thame (Oxon.); through Long Crendon, Chilton, Oakley, Boarstall (Bucks.); Ambrosden, Arncot, Merton, Bicester, Bucknell, Ardley, Fritwell, Souldern (Oxon.); Aynho (Northants.); terminating there in a junction with the Oxford & Rugby Junction Railway.

33. Rugby & Huntingdon Railway, with a branch by Market Harborough to Stamford. No engineer named. Deposited 30 Nov. 1845. No Act.

Main Line. From a junction with the London & Birmingham Railway near Rugby Station through Rugby, Clifton on Dunsmore (Warwicks.); Lilbourne (Northants.); Catthorpe, Swinford (Leics.); Stanford, Welford (Northants.); Husbands Bosworth (Leics.); Sulby, Naseby, Clipston, Great Oxendon, Kelmarsh, Arthingworth, Harrington, Desborough, Rothwell, Thorpe Malsor, Cransley, Kettering, Broughton, Pytchley, Barton Seagrave, Burton Latimer, Finedon, Great Addington, Woodford near Thrapston, Ringstead, Raunds (Northants.); Keyston, Bythorn, Molesworth, Brington, Catworth, Leighton Bromswold, Spaldwick, Ellington, Alconbury, Little Stukeley, Great Stukeley, Huntingdon (Hunts.); terminating in a junction with the London & York Railway.

Branches from the main line to join the Northampton & Peterborough Railway (i.e. the Blisworth & Peterborough Branch of the London & Birmingham Railway). Through Great Addington, Woodford near Thrapston, Ringstead, Raunds (Northants.).

Stamford and Market Harborough Branch. From a junction with the London & York Railway in Stamford (Lincs.); through Tinwell (Rutland); Easton on the Hill (Northants.); Ketton (Rutland); Collyweston, Duddington (Northants.); Tixover (Rutland); Wakerley (Northants.); Barrowden, Seaton (Rutland); Harringworth (Northants.); Thorpe by Water, Liddington (Rutland); Gretton (Northants.); Caldecott (Rutland); Great Easton, Bringhurst (Leics.); Cottingham, East Carlton, Ashley (Northants.); Medbourne, Slawston (Leics.); Weston by Welland (Northants.); Thorpe Langton, Great Bowden (Leics.); Little Bowden, Great Oxendon, Braybrooke, Clipston (Northants.); terminating in junctions with the Leicester & Bedford Railway and the main line as above.

34. East & West of England Junction Railway. J.B. & E. Birch, engineers. Deposited 30 Nov. 1845. No Act.

From a junction with the London & Birmingham Railway at Blisworth Station, through Blisworth, Tiffield, Towcester, Abthorpe, Wappenham, Weedon Lois, Helmdon, Sulgrave, Culworth, Thorpe Mandeville, Edgcote (Northants.); Wardington (Oxon.); Chalcombe, Warkworth, Grimsbury, Easton Neston (Northants.); Banbury, Neithrop, Broughton, North Newington, Tadmarton, Bloxham, Milcombe, Wigginton, South Newington, Swerford, Hook Norton, Great Rollright, Over Norton, Chipping Norton, Churchill, Sarsden, Kingham, Heythrop (Oxon.); terminating in a junction with the Oxford, Worcester & Wolverhampton Railway.

35. Peterborough, Spalding & Boston Junction Railway *(thus on cover of plan and section and book of reference;* **Peterborough, Wisbech, Lynn & Boston Junction Railway** *on first sheet of plan and section).* No engineer named. Deposited 30 Nov. 1845. No Act.

From a junction with the Eastern Counties Railway at Peterborough Station in Fletton (Hunts.), through Stanground (Cambs. & Hunts.); Peterborough, Eye, Newborough, Borough Fen (Northants.); Crowland, Deeping Fen, Pinchbeck, Spalding (Lincs.); terminating at the turnpike road from Spalding to Bourn.

36. London, Warwick, Leamington & Kidderminster Railway, with branch to the Birmingham & Gloucester Railway at Coston *(sic)* **Hackett.** Robert Stephenson, consulting engineer. John Addison, engineer. Deposited 29 Nov. 1845. No Act.

Main Line. From a junction with the London & Birmingham Railway in Weedon Beck, through Everdon, Newnham, Badby, Staverton (Northants.); Upper Shuckburgh, Wolfhampcote, Lower Shuckburgh, Grandborough, Caldecote, Napton on the Hill, Stockton, Southam, Long Itchington, Ufton, Radford Semele, Leamington Priors, Warwick, Budbrooke, Sherborne, Norton Lindsey, Fulbrook, Snitterfield, Wolverton, Claverdon, Langley, Wootton Wawen, Bearley, Aston Cantlow, Great Alne, Coughton, Sambourne, Spernall (Warwicks.); Feckenham, Hanbury, Tardebigge, Stoke Prior, Bromsgrove, Grafton Manor, Dodderhill, Elmbridge, Upton Warren, Chaddesley Corbett, Stone, Kidderminster; terminating in a junction with the Welsh Midland Railway (Worcs.)

Branch. From a junction with the main line in Sambourn, through Studley (Warwicks.); Ipsley, Tardebigge, Redditch,

Tutnall & Cobley, Alvechurch, Cofton Hackett (Worcs.); terminating in a junction with the Birmingham & Gloucester Railway.

37. Rugby & Stamford Railway. Robert Stephenson, Charles Liddell, engineers. Deposited 30 Nov. 1845. No Act.

From a junction with the London & Birmingham Railway near Rugby Station, through Rugby, Clifton on Dunsmore (Warwicks.); Lilbourne (Northants.); Catthorpe, Swinford (Leics.); Stanford, Welford (Northants.); South Kilworth, North Kilworth, Husbands Bosworth, Theddingworth, Lubenham (Leics.); Thorpe Lubenham, East Farndon, Little Bowden (Northants.); Great Bowden (Leics.); Church Langton (Leics.); Weston by Welland (Northants.); Medbourne (Leics.); Ashley (Northants.); Bringhurst (Leics.); Cottingham, East Carlton (Northants.); Great Easton (Leics.); Caldecott, Liddington, Thorpe by Water, Seaton, Morcott, South Luffenham (Rutland); terminating in a junction with the Syston & Peterborough Railway.

38. Buckinghamshire Railway. Tring and Banbury Line. No engineer named. Deposited 29 Nov. 1845. Act 9 & 10 Vict. c ccxxxiii (1846).

From a junction with the London & Birmingham Railway at Tring Station, through Aldbury, Tring (Herts.); Drayton Beauchamp, Buckland, Aston Clinton, Weston Turville, Bierton with Broughton, Aylesbury, Quarrendon, Fleet Marston, Waddesdon, Pitchcott, Quainton, Hogshaw, North Marston, Grandborough, East Claydon, Middle Claydon, Steeple Claydon, Padbury, Buckingham, Radclive, Tingewick, Water Stratford (Bucks.); Finmere (Oxon.); Westbury (Bucks.); Mixbury (Oxon.); Evenley (Northants.); Turweston (Bucks.); Brackley St Peter, Brackley St James (enlarged plan of part of town), Hinton in the Hedges, Steane, Farthinghoe, Greatworth, Marston St Lawrence, Thenford, Middleton Cheney, King's Sutton, Newbottle, Warkworth (Northants.); terminating at Banbury (enlarged plan of part of town) (Oxon.).

39. Midland & Eastern Counties Railway. John Miller, engineer. Deposited 29 Nov. 1845. No Act.

From a junction with the Northern & Eastern Railway, or Cambridge Line of the Eastern Counties Railway, in Cambridge, through Trumpington, Grantchester, Barton, Haslingfield, Harlton, Little Eversden, Comberton, Great Eversden, Toft, Kingston, Caldecote, Bourn, Caxton (Cambs.); Great Gransden (Hunts.); Croxton (Cambs.); Abbotsley, Eynesbury, St Neots (Hunts.); Eaton Socon, Colmworth, Bolnhurst, Thurleigh, Keysoe, Riseley, Bletsoe, Sharnbrook, Knotting, Souldrop, Podington (Beds.); Wollaston, Bozeat, Strixton, Grendon, Castle Ashby, Whiston, Cogenhoe, Ecton, Great Billing, Little Billing, Brafield on the Green, Weston Favell, Abington, Northampton St Giles, Northampton All Saints (enlarged plan of Bridge Street), Hardingstone, Duston, Upton, Kislingbury, Bugbrooke, Nether Heyford, Stowe Nine Churches, Weedon Beck (Northants.); terminating in a junction with the London & Birmingham Railway.

40. Peterborough & Nottingham Junction Railway. Sir John Rennie and George Rennie, consulting engineers. William Lewin, acting engineer. Deposited 30 Nov. 1845. No Act.

Part I. Stamford to Nottingham. From a junction with the Midland Counties Railway in Stamford Baron (Northants.); through Wothorpe (Northants.); Stamford (Lincs.); Little Casterton, Great Casterton, Tickencote, Empingham, Greetham, Cottesmore, Barrow, Market Overton (Rutland); Edmondthorpe, Wymondham, Garthorpe, Saxby, Freeby, Melton Mowbray, Thorpe Arnold, Brentingby & Wyfordby, Sysonby, Holwell, Ab Kettleby, Nether Broughton (Leics.); Hickling, Kinoulton, Owthorpe, Cropwell Bishop, Cotgrave, Tollerton, Holme Pierrepont, Gamston, West Bridgford, South Wilford, Nottingham (Notts.); terminating there in a junction with the Midland Counties Railway.

Part II. Melton Mowbray to the Main Line. Wholly in Melton Mowbray.

41. Midland Railway. Branches and Deviations from the Syston & Peterborough Railway. Robert Stephenson, Charles Liddell, engineers. J.G. Binns, surveyor. Deposited 30 Nov. 1845. Act 9 & 10 Vict. c. li.

Branch at Syston. From a junction with the Midland Railway near Syston Station to the the parliamentary line, all in Syston (Leics.).

Deviation from Brooksby to Frisby on the Wreak. In Brooksby, Hoby, Rotherby, Frisby on the Wreak (Leics.).

Northern Deviation from Melton Mowbray. In Melton Mowbray, Thorpe Arnold, Brentingby & Wyfordby, Saxby, Wymondham, Edmondthorpe (Leics.); Whissendine (Rutland).

Southern Deviation from Melton Mowbray. In Melton Mowbray, Thorpe Arnold, Brentingby & Wyfordby, Stapleford (Leics.); Whissendine, Teigh, Ashwell (Rutland).

Deviation at Oakham. In Burley, Oakham (Rutland).

Deviation from Easton to Stamford by Ufford. In Easton on the Hill (Northants.); Stamford (Lincs.); Stamford Baron, Barnack, Ufford, Bainton (Northants.).

Branch from Barnack to Elton. From a junction with the Syston & Peterborough Railway at Barnack, through Ufford, Thornhaugh, Wansford (Northants.); Sibson cum Stibbington (Hunts.); Yarwell (Northants.); Elton (Hunts.); terminating in a junction with the Northampton & Peterborough Railway.

42. South Midland, or Leicester, Northampton, Bedford & Huntingdon Railways. Robert Stephenson, Charles Liddell, engineers. Plan and section endorsed with memorial dated 30 Nov. 1845 (signed by Charles Markham); book of reference (which corresponds with plan) bears printed date Nov. 1846 and memorial (signed by H.P. Markham) dated 30 Nov. 1846. Act 10 & 11 Vict. c. cxxxv (1847).

Main Line. From junction with Midland Railway near Wigston Station, through Wigston Magna, Wistow, Burton Overy, Kibworth Beauchamp, Church Langton, Foxton, Great Bowden (Leics.), Little Bowden, Braybrooke, Desborough, Rushton, Barford, Glendon, Kettering, Pytchley, Isham, Little Harrowden, Great Harrowden, Wellingborough, Irthlingborough, Irchester (Northants.); Wymington, Souldrop, Sharnbrook, Felmersham, Milton Ernest, Pavenham, Oakley, Bromham, Biddenham, Bedford St Paul, Bedford St John, Bedford St Mary (Beds.); terminating in a junction with existing railway. Not planned beyond junction but book of reference continues as follows: Elstow, Cardington, Old Warden, Southill, Shefford Hardwick, Clifton, Henlow, Arlesey (Beds.); Holwell, Ickleford, Hitchin (Herts.).

Line from Little Bowden to Northampton. From junction with main line in Little Bowden, through Great Oxenden, Kelmarsh, Arthingworth, Maidwell, Draughton, Lamport, Cottesbrooke, Great Creaton, Brixworth, Spratton, Church Brampton, Pitsford, Boughton, Kingsthorpe, Dallington, Northampton St Sepulchre, Northampton St Peter, Northampton All Saints, Duston, Hardingstone (Northants.); terminating in a junction with the Northampton & Peterborough Branch of the London & North Western Railway.

Line to Huntingdon. From junction with main line in Isham, through Burton Latimer, Finedon, Great Addington, Little Addington, Raunds (Northants.); Keyston, Bythorn, Molesworth, Catworth, Leighton Bromswold, Spaldwick, Ellington, Easton, Alconbury, Little Stukeley, Great Stukeley, Huntingdon St John, Huntingdon St Mary, Brampton, Godmanchester (Hunts.); terminating in a junction with the Ely & Huntingdon Railway.

Branch to Wellingborough and branch therefrom to main line. From junction with main line in Wellingborough, through Irchester (Northants.).

Branch from Raunds to Irthlingborough. From terminus in Irthlingborough, through Stanwick, Raunds (Northants.); terminating in junction with main line.

Proposed road to the intended station at Market Harborough. Land in Market Harborough parish (Leics.) (included in book of reference but not planned).

Enlargement of the Leicester Station. Land in Leicester St Margaret (Leics.) (included in book of reference but not planned).

43. London & Nottingham Railway. C.H. Gregory, engineer-in-chief. Thomas Hawkesley, Nathaniel Briant, engineers. Deposited 30 Nov. 1845. No Act.

From a junction with the Midland Railway near that company's Nottingham Station, through Nottingham St Mary, South Wilford, West Bridgford, Gamston, Holme Pierrepont, Tollerton, Cotgrave, Cropwell Bishop, Colston Basset (Notts.); Long Clawson, Hose, Melton Mowbray, Thorpe Arnold, Burton Lazars, Stapleford (Leics.); Whissendine, Langham, Barleythorpe, Oakham, Manton, Wing, Bisbrooke, Seaton, Liddington, Thorpe by Water (Rutland); Gretton, Corby, Great Oakley, Little Oakley, Newton, Geddington, Weekley, Kettering, Pytchley, Barton Seagrave, Burton Latimer, Isham, Little Harrowden, Great Harrowden, Wellingborough, Irthlingborough, Irchester (Northants.); Wymington, Podington, Souldrop, Sharnbrook, Bletsoe, Milton Ernest, Oakley, Clapham, Bedford (Beds.); terminating at Bedford.

44. Port of Wisbech, Peterborough, Birmingham & Midland Counties Union Railway. B. Albano, W. Laxton, engineers. Coe & Mann, surveyors. Deposited 30 Nov. 1845. No Act.

Main Line. From a junction with the proposed Lynn & Ely Railway in Wisbech, through Thorney (Cambs.); Eye, Peterborough (Northants.); Fletton, Woodstone (Hunts.), terminating there in a junction with the Northampton & Peterborough branch of the London & North Western Railway.

Branch Line. From the main line to the river Nene, all in Wisbech (Cambs.).

45. Lynn, Wisbeach & Peterborough, Midland Counties & Birmingham Junction Railway. Thomas Rumball, engineer. Deposited 12.05 a.m. 1 Dec. 1845. No Act.

Main Line. From a junction with the Northampton & Peterborough Railway in Fletton (Hunts.), through Stanground (Cambs. & Hunts.); Peterborough (Northants.), Whittlesey St Mary & St Andrew, Thorney, Wisbech (enlarged plan of town) (Cambs.); Walsoken, West Walton, Walpole St Peter, Terrington St John, Tilney St Lawrence, Tilney with Islington, Tilney All Saints, St Mary All Saints, Clenchwarton, Wiggenhall St Mary the Virgin (Norfolk); terminating in a junction with the intended Lynn & Ely Railway.

Branch Line. From the main line in Walsoken (Norfolk) to the river Nene in Wisbech (Cambs.).

Alternative Main Line. From a junction with the Eastern Counties Railway in Stanground (Cambs. & Hunts.); through Whittlesey St Mary & St Andrew, Wisbech, Elm (Cambs.), rejoining the main line near Wisbech.

46. Peterborough, Wisbech & Lynn Junction Railway. H.E. Scott, engineer. Frederick J. Utting, surveyor. Deposited 30 Nov. 1845. No Act.

From a junction with the Eastern Counties Railway in Fletton (Hunts.), through Stanground (Cambs. & Hunts.); Peterborough, Eye (Northants.); Thorney, Sutton St Edmund, Parson Drove, Leverington, Wisbech (Cambs.), terminating at Wisbech in two spurs, one running to the river Nene in the town and the other to the Nene at Leverington.

47. Northampton, Banbury & Cheltenham Railway. No engineer named. Deposited 30 Nov. 1845. No Act.

Two separate sets of plans and sections, titled as below and listed as 47 and 47a; no book of reference.

47. From Northampton to Ashchurch. From a junction with the London & Birmingham Railway in Blisworth, through Gayton, Tiffield, Towcester, Bradden, Abthorpe, Slapton, Wappenham, Astwell with Falcutt, Helmdon, Stuchbury, Greatworth, Farthinghoe, Marston St Lawrence, Thenford, Middleton Cheney, King's Sutton, Newbottle, Warkworth (Northants.); Banbury, Neithrop, North Newington, Broughton, Bloxham, Tadmarton, Swalcliffe, Hook Norton, Sibford Ferris, Sibford Gower (Oxon.); Whichford, Stourton, Brailes, Sutton under Brailes, Cherington, Burmington (Warwicks.); Todenham (Gloucs.); Tidmington, Stretton on Fosse (Warwicks.); Blockley, Ebrington, Mickleton, Weston Subedge, Aston Subedge, Saintbury, Willersey (Gloucs.); Broadway (Worcs.); Buckland, Wormington (Gloucs.); Aston Somerville (Worcs.); Dumbleton (Gloucs.); Ashton Underhill, Beckford, Overbury (Worcs.); Ashchurch (Gloucs.), terminating in a junction wth the Birmingham & Gloucester Railway.

47a. Northampton to Banbury. One sheet of plans, which is a duplicate of the first sheet of the preceding set.

48. London & Birmingham Railway. Extensions at Birmingham; at Leamington; from Weedon to Northampton; at Coventry; at Rugby. Deposited 30 Nov. 1845. Act 9 & 10 Vict. c. cccix (1846).

Extension at Birmingham. Robert Stephenson, William Baker, engineers. In Aston, Birmingham (Warwicks.). Large-scale plan of town centre and exceptionally detailed book of reference.

Extension at Leamington. Robert Stephenson, engineer. In Milverton, Warwick Leamington Priors (Warwicks.). Enlarged plan showing new line between existing station on Warwick to Leamington road and town centre.

Extension from Weedon to Northampton. Robert Stephenson, engineer. From junction with existing main line at Weedon Beck, through Stowe Nine Churches, Nether Heyford, Bugbrooke, Kislingbury, Wootton, Hardingstone (Northants.); terminating in junction with Northampton & Peterborough Railway.

Extension at Coventry. No engineer named. Land in Coventry (Warwicks.).

Extension at Rugby. No engineer named. Land in Rugby (Warwicks.).

49. Eastern Counties Railway. York Extension. Line between Cambridge and Lincoln, with branches to Sleaford, Boston and Spalding. J.M. Rendel, engineer. Lister & Mills, surveyors. Deposited 30 Nov. 1845. No Act.

Main Line. From junction with Cambridge & Ely Railway near Cambridge Station, through Cambridge St Andrew the Less, Chesterton, Girton, Impington, Histon, Oakington, Long Stanton St Michael, Long Stanton All Saints, Swavesey, Fen Drayton (Cambs.); Fen Stanton, Holywell cum Needingworth, St Ives, Woodhurst, Old Hurst, Broughton, Warboys, Wistow, Bury, Ramsey, Farcet (Hunts.); Stanground (Cambs. & Hunts.); Fletton (Hunts.); Peterborough, Paston, Glinton, Etton, Northborough, Maxey (Northants.); Market Deeping, Langtoft, Baston, Thurlby, Bourne, Morton, Hacconby, Dunsby, Rippingale, Dowsby, Aslackby, Sempringham, Pointon, Billingborough, Horbling, Swaton, Helpringham, Great Hale, Little Hale, Heckington, Asgarby, Ewerby, Haverholme Priory, Anwick, Ruskington, Dorrington, Digby, Rowlston, Timberland, Martin, Thorpe Tilney, Scopwick, Blankney, Linwood, Metheringham, Dunston, Nocton, Potter Hanworth, Branston, Washingborough, Heighington, Canwick, Lincoln (Lincs.); terminating at a junction with the intended Nottingham & Lincoln Railway and proposed continuation line to York.

Sleaford Branch. From a terminus at Sleaford, through Quarrington, Old Sleaford, Kirkby la Thorpe, Heckington (Lincs.); terminating in a junction with the main line.

Boston Branch. From a junction with the main line in Heckington, through Great Hale, Swineshead, Algarkirk, Sutterton, Kirton, Frampton, Wyberton, Skirbeck, Boston (Lincs.); terminating there.

Spalding Branch, with continuation to Boston. From a junction with the main line in Market Deeping, through Langtoft, Greatford, Barholm, Tallington, Uffington, Deeping Fen, Spalding, Pinchbeck, Surfleet, Gosberton, Sutterton, Algarkirk, Kirton, Frampton, Wyberton, Skirbeck, Boston (Lincs.); terminating there.

Alternative Line at St Ives. In Fen Drayton (Cambs.); Fen Stanton, St Ives (Hunts.).

50. London & Birmingham Extension, Northampton, Daventry, Leamington & Warwick Railway, with branches to Rugby and Fenny Compton. Sir John Macneill and James Thomson, consulting engineers. William B. Prichard, engineer. Deposited 30 Nov. 1845. No Act.

From a terminus in Bridge Street, Northampton (enlarged plan), through Duston, Hardingstone, Upton, Kislingbury, Harpole, Upper Heyford, Flore, Stowe Nine Churches, Dodford (junction with the London & Birmingham main line), Newnham, Norton, Daventry (enlarged plan), Staverton, Braunston (Northants.); Wolfhampcote, Lower Shuckburgh, Grandborough, Napton on the Hill, Leamington Hastings, Birdingbury, Stockton, Long Itchington, Offchurch, Radford Semele, Leamington Priors, Warwick (Warwicks.); terminating at Warwick in junctions with the Warwick & Birmingham Canal Railway and Warwick & Worcester Railway.

Alternative Main Line: Through Grandborough, Napton, Southam (enlarged plan), Long Itchington, Ufton, Offchurch, Radford Semele, Leamington Priors (enlarged plan of town), Milverton, Warwick (Warwicks.).

Rugby Branch: From main line in Stockton, through Birdingbury and Hastings (Warwicks.), terminating in a junction with intended Oxford & Rugby Railway.

Fenny Compton Branch: From main line in Leamington Priors, through Whitnash, Radford Semele, Harbury, Bishops Itchington, Burton Dassett (Warwicks.), terminating in a junction with the intended Oxford & Rugby Railway.

51. Midland Grand Junction Railway; extending from the town of Reading, Berkshire, to the station at Blisworth, Northamptonshire. [Deposited 29–30 Nov. 1845.] No Act.

An exceptionally crude series of plans and sections, some marked 'Proof', on which the basic survey lines are lithographed but most legend completed in ink, with three lithographed cover sheets, each signed 'Oliver Byrne Engineer', two in ink and one in pencil; numerous cover sheets for books of reference but no actual schedules; drafts of schedules for Whittlebury (Northants.) and Lillingstone Lovell (Bucks.), on printed forms but with details completed in ink, of which the latter is endorsed 'Not wanted. Copied in other references'; two prints of extracts from House of Lords Standing Orders, one endorsed 'Recd Novr. 29 1845 J.M.' and other 'Recd Novr 30 1845 J.M.'. All these items were once enclosed in a cardboard portfolio, lacking a deposit memorial or other endorsement. The plans do not form a complete set covering the route from Reading to Blisworth and there is more than one print of some sheets.

Two copies of the book of reference, deposited 30 Nov. 1845. Almost all completed in ink on printed forms. From Great Western Railway near Reading Station at Reading (Berks.); through Caversham, Eye & Dunsden, Shiplake, Harpsden (Oxon.); Wargrave, Remenham, Hurley (Berks.); Medmenham, Great Marlow, Chepping Wycombe, West Wycombe, Bradenham, Saunderton, Horsendon, Princes Risborough, Monks Risborough, Great Kimble, Stone, Aylesbury, Quarrendon, Bierton with Broughton, Waddesdon, Pitchcott, Quainton, North Marston, Hogshaw, East Claydon, Middle Claydon, Addington, Adstock, Padbury, Buckingham, Maids Moreton, Foscott, Akeley, Leckhampstead, Lillingstone Lovell (Bucks.); Potterspury, Passenham, Whittlebury, Yardley Gobion, Paulerspury, Alderton, Stoke Bruerne, Shutlanger, Easton Neston, Tiffield, Blisworth (Northants.); terminating in a junction with the London & Birmingham Railway.

52. Great Northern Railway. Deviations between London and Grantham. William Cubitt, consulting engineer. Joseph Cubitt, engineer. Sherrard & Hall, surveyors. Deposited 28 Nov. 1846. Act 10 & 11 Vict. c. cclxxxvii (1847).

King's Cross Deviations: Main Line. In Islington St Mary and St Pancras (Middlesex).

Levels Alteration. In Hornsey, Tottenham, Edmonton, Friern Barnet, East Barnet, Monken Hadley, Enfield, South Mimms (Middlesex).

Hatfield Deviation. In North Mimms (Middlesex) and Hatfield (Herts.).

Peterborough Deviation: Main Line. In Fletton (Hunts.); Peterborough, Paston, Marholm, Glinton, Etton, Helpston, Ufford, Bainton, Barnack, Pilsgate (Northants.); Uffington (Lincs.); Stamford Baron (Northants.); Stamford (Lincs.); Ryhall, Essendine (Rutland); Carlby, Holywell, Careby (Lincs.).

Stamford and Spalding. Western Extension. In Ufford, Bainton and Maxey (Northants.).

Grantham Deviation. In Careby, Little Bytham, Creeton, Castle Bytham, Counthorpe, Swayfield, Corby, Burton Coggles, Bassingthorpe, South Stoke, Great Ponton, Little Ponton, Grantham, Great Gonerby (Lincs.).

53. Buckingham & Brackley and Oxford & Bletchley Junction Railway. Extensions to Banbury and Aylesbury and into Oxford. Robert Stephenson, Robert B. Dockray, engineers. C.F. Cheffins, lithographer. Deposited 30 Nov. 1846. Act 9 & 10 Vict. ccxxxvi (1847).

Banbury Extension. From a junction with the intended Buckingham & Brackley Junction Railway in Brackley St James; through Brackley St Peter, Steane, Greatworth, Farthinghoe, Middleton Cheney, Marston St Lawrence, Warkworth, Grimsbury (Northants.); terminating there in a junction with the Oxford and Rugby Railway.

Aylesbury Extension. From a junction with the LNWR in Aylesbury; through Bierton with Broughton, Waddesdon, Pitchcott, Quainton, North Marston, Grandborough, East Claydon, Middle Claydon (Bucks.); terminating there in a junction with the intended line of the Oxford & Bletchley Railway.

Oxford Extension. From a junction with the intended Oxford & Bletchley Railway in Cutteslowe, through Kidlington, terminating in Oxford St Giles in a junction with the intended London, Oxford & Cheltenham Railway (Oxon.).

54. Boston, Stamford & Birmingham Railway. Peterborough & Thorney Line. Commencing by a junction with the Syston & Peterborough Railway in St John the Baptist in or near the city of Peterborough (Northants.), and terminating by a junction with the Stamford & Wisbech Line of the Boston, Stamford & Birmingham Railway, as authorised by the Boston, Stamford & Birmingham Railway Act 1846. Stamford & Wisbech Line in Thorney, in the Isle of Ely (Cambs.). Deposited 28 Nov. 1846. No engineer or surveyor named. Act 10 & 11 Vict. c. cxii (1847).

Peterborough and Thorney Line. From junction with Syston & Peterborough Railway in Peterborough, through Newborough, Eye (Northants.); Thorney (Cambs.), terminating there in a junction with the Boston, Stamford & Birmingham Railway (Stamford & Wishbech Line). Index map also shows route of latter from junction with Syston & Peterborough Railway at Lolham Bridges (Northants.) to Wisbech (Cambs.); a Harbour Branch at Wisbech; and a line from Wisbech to Sutton (Lincs.), but these are not included in the plan or book of reference deposited in Northamptonshire.

55. Northampton & Banbury Railway. Cover sheet of plan largely torn away, possibly removing longer title, name of engineer etc. Deposited 30 Nov. 1846. Act 10 & 11 Vict. clxxviii (1847).

From a junction with the London & North Western Railway near Blisworth Station, through Gayton, Tiffield, Towcester, Easton Neston, Bradden, Slapton, Abthorpe, Wappenham, Weedon Lois, Astwell with Falcutt, Helmdon, Stuchbury, Greatworth, Farthinghoe, Middleton Cheney, Marston St Lawrence, Warkworth and Grimsbury (Northants.); terminating in a junction with the Oxford & Rugby Railway.

56. Midland Railway Extensions. From Leicester to Hitchin and to Northampton and Huntingdon. Robert Stephenson, Charles Liddell, engineers. J.G. Binns, surveyor. Deposited 30 Nov. 1845. Act 10 & 11 Vict. c. cxxxv (1847).

Northampton Line. From the company's line at Wigston Magna, through Wistow, Glen Magna, Burton Overy, Kibworth, Beauchamp, Church Langton, Foxton, Great Bowden (Leics.); Little Bowden, Great Oxendon, Kelmarsh, Arthingworth, Maidwell, Draughten, Lamport, Cottesbrooke, Brixworth, Spratton, Church Brampton, Pitsford, Boughton, Kingsthorpe, Dallington, Northampton, Duston, Hardingstone (Northants.).

Bedford Line. From a junction with the Northampton line at Little Bowden, through Braybrooke, Desborough, Rushton, Barford, Glendon, Kettering, Pytchley, Barton Seagrave, Burton Latimer, Isham, Little Harrowden, Finedon, Great Harrowden, Wellingborough, Irthlingborough, Irchester (Northants.); Wymington, Souldrop, Sharnbrook, Felmersham, Milton Ernest, Pavenham, Oakley, Bromham, Biddenham, Bedford (Beds.).

Branch to Huntingdon. Through Pytchley, Burton Latimer, Finedon, Great Addington, Little Addington, Raunds (Northants.); Keystone, Bythorn, Molesworth, Brington, Catworth, Leighton Bromswold, Spaldwick, Ellington, Easton, Alconbury, Little Stukeley, Great Stukeley, Huntingdon, Brampton, Godmanchester (Hunts.).

Branch from the Huntingdon Line to join the Northampton & Peterborough Railway in Irthlingborough. In Raunds, Stanwick and Irthlingborough (Northants.).

Branch from the Bedford Line to join the Northampton & Peterborough Railway in Irchester. In Irchester.

57. Drainage of Crowland Cowbit Washes and Other Lands. Cartwright & Pritchard, lithographers. Deposited 27 Nov. 1846. Act 10 & 11 Vict. c. cclxvii (1847).

Lands in Maxey, Newborough (Northants.); Deeping St James, Crowland, Deeping Fen, Pinchbeck, Cowbit & Peakill, Spalding (Lincs.).

58. Midland Railway. Syston & Peterborough Railway. Deviations and approach to Manton Station. No engineer named. Deposited 30 Nov. 1846. Act 10 & 11 Vict. c. ccxv (1847).

Deviation from Melton Mowbray to Ashwell. In Melton Mowbray, Thorpe Arnold, Brentingby & Wyfordby, Saxby, Wymondham, Edmondthorpe (Leics.); Whissendine, Teigh, Ashwell (Rutland).

Deviation from Barnack to the Northampton & Peterborough line. In Barnack and Castor (Northants.).

Approach to Manton Station. In Manton (Rutland).

59. Great Northern Railway. Deviations between Peterborough, Doncaster and Boston. Deposited 28 Nov. 1846. No Act.

Peterborough to Spalding Deviation. William Cubitt, consulting engineer. John Miller, engineer. Samuel Hughes, surveyor. From Fletton (Hunts.), through Peterborough, Eye, Newborough, Borough Fen (Northants.); Crowland, Deeping Fen, Spalding, Cowbit (Lincs.).

Stamford and Spalding Branch Eastern Extension. William Cubitt, consulting engineer. John Miller, engineer. Samuel Hughes, surveyor. From Deeping St James, through Deeping Fen and Spalding (Lincs.).

Boston Deviation. William Cubitt, consulting engineer. Joseph Cubitt, engineer. From Skirbeck to Boston (Lincs.).

Doncaster and Gainsborough Deviation. William Cubitt, consulting engineer. John Miller, engineer. From Saundby, through Beckingham, Walkeringham, Misterton, Gringley on the Hill (Notts.); Haxey (Lincs.); Misson, Blyth, Austerfield (Notts.); Finningley (Notts.); Rossington, Cantley, Doncaster, Balby with Hexthorpe, Bentley with Arksey (Yorks.).

60. Eastern Counties Railway. Peterborough to Folkingham. Robert Stephenson, M.A. Borthwick, engineers. C.F. Cheffins, surveyor. Deposited 30 Nov. 1846.

From a junction with the Syston & Peterborough Railway of the Midland Counties Railway in Etton, through Maxey (Northants.); Market Deeping, Langtoft, Baston, Thurlby, Bourne, Morton, Hacconby, Dunsby, Rippingale, Dowsby, Aslackby, Pointon, Sempringham, Laughton, Folkingham (Lincs.); terminating in a junction with the Ambergate, Nottingham & Boston Railway.

61. London & North Western Railway. Newport Pagnell, Olney Branch. Robert Stephenson, H.E. Scott, engineers. C.F.

Cheffins, surveyor. Deposited 28 Nov. 1846. Act 10 & 11 Vict. c. cvii (1847).

From a junction with the LNWR main line at Bletchley, through Bletchley, Fenny Stratford, Woughton on the Green, Great Woolstone, Little Woolstone, Willen, Newport Pagnell, Lathbury, Sherington, Tyringham with Filgrave, Emberton, Olney, Warrington (Bucks.); Bozeat, Easton Maudit, Grendon, Strixton, Wollaston, Irchester, Great Doddington, Wellingborough (Northants.); terminating in a junction with the Northampton & Peterborough branch of the LNWR.

62. River Nene. Improvement and Drainage. Deposited 30 Nov. 1847. Act 11 & 12 Vict. c. cxliii (1848).

Includes lands in Long Sutton (Lincs.); Walpole St Andrew, Walpole St Peter (Norfolk); Tydd St Mary (Lincs.); Tydd St Giles (Cambs.); West Walton, Walsoken (Norfolk); Wisbech, Leverington, Whittlesey St Mary & St Andrew (Cambs.); Stanground (Cambs. & Hunts.); Fletton, Woodstone (Hunts.); Peterborough (Northants.).

63. Midland Railway. Alteration of the line and branches near Wellingborough and station approaches. Deposited 30 Nov. 1847. Act 11 & 12 Vict. c. xxi (1848).

Alteration near Wellingborough. In Isham, Little Harrowden, Great Harrowden, Finedon, Wellingborough, Irthlingborough, Irchester (Northants.).

Station approaches. In Wellingborough (Northants.).

64. Nene Valley Drainage and Navigation Improvement. James Meadows Rendel, engineer in chief. William Radford, acting engineer. Deposited 29 Nov. 1851. Act 15 & 16 Vict. c. cxxviii (1852), amended by 17 & 18 Vict. c. lxxxii (1854).

Deposit wrapper also includes (a) volume listing owners and occupiers of lands subject to taxation under the Act, with parcel numbers from the deposited plan, and tax assessed on each parcel, deposited 31 July 1854; (b) lists of names of commissioners and deeds of appointment of commissioners, enrolled with the clerk of the peace, 1777, 1781, 1785, 1825, 1827, 1830, 1837 and 1842.

Upper Division: Works in Kislingbury, Upton, Duston, Wootton, Chapel Brampton, Church Brampton, Pitsford, Boughton, Kingsthorpe, Dallington, Northampton, Far Cotton, Cotton End, Hardingstone, Abington, Great Houghton, Little Houghton, Weston Favell, Little Billing, Great Billing, Brafield on the Green, Denton, Cogenhope, Ecton, Earls Barton, Whiston, Castle Ashby, Grendon, Strixton, Wilby, Great Doddington, Wollaston, Wellingborough, Irchester, Rushden, Irthlingborough, Higham Ferrers, Chelveston cum Caldecott, Stanwick, Little Addington, Raunds, Ringstead, Great Addington, Denford, Woodford near Thrapston, Twywell, Islip, Thrapston, Aldwincle, Titchmarsh, Thorpe Achurch, Wadenhoe, Pilton, Lilford cum Wigsthorpe, Stoke Doyle, Barnwell St Andrews, Barnwell All Saints, Polebrook, Armston, Oundle, Ashton [by Oundle], Glapthorn, Cotterstock, Tansor, Southwick, Warmington, Fotheringhay, Woodnewton, Apethorpe, Nassington, Yarwell, Wansford, Thornhaugh, Sutton, Wittering, Ailsworth, Castor, Peterborough, Long Thorpe (Northants.); Elton, Sibson cum Stibbington, Water Newton, Chesterton, Alwalton, Orton Waterville, Orton Longville, Woodstone, Fletton (Hunts.); Stanground (Cambs. & Hunts.); Whittlesey St Mary & St Andrew, Coates, Guyhirn, Elm, Parson Drove (Cambs.); Emneth (Norfolk); Leverington, Thorney, Wisbech (Cambs.).

Lower Division: Works in Wisbech and Leverington (Cambs.).

Moreton's Leam: Works in Fletton (Hunts.); Stanground (Cambs. & Hunts.); Whittlesey St Mary & St Andrew, Eldernell, Coates, Estrea, Guyhirn, Elm, Wisbech (Cambs.).

Smith's Leam: Works in Wisbech, Guyhirn, Elm, Whittlesey St Mary & St Andrew, Eldernell, Coates, Estrea, Stanground (Cambs. & Hunts.); Fletton, Woodstone (Hunts.); Longthorpe, Peterborough (Northants.).

65. Stamford & Essendine Railway, with a branch to join the Midland Railway at Stamford, and works for the improvement of the river Welland at or near Stamford. William Hurst, surveyor. Deposited 30 Nov. 1852. Act 16 & 17 Vict. c. cxcix (1853).

Railway. From a junction in Stamford Baron (Northants.) with a branch to the Syston & Peterborough branch of the Midland Railway, through Stamford, Uffington (Lincs.); Ryhall, Essendine (Rutland); terminating in a junction with the Great Northern Railway.

Works for the improvement of the river Welland. In Stamford (Lincs.); Stamford Baron (Northants.)

66. London & North Western Railway. Branch from Northampton to Market Harborough. George P. Bidder, George Robert Stephenson, engineers. R. Apsley Ranger, surveyor. Deposited 30 Nov. 1852. Act 16 & 17 Vict. c. clx (1853).

From a junction with the Northampton & Peterborough Branch of the LNWR in Hardingstone, through Far Cotton, Duston, Northampton, Upton, Dallington, Kingsthorpe, Boughton, Church Brampton, Chapel Brampton, Pitsford, Brixworth, Spratton, Great Creaton, Cottesbrooke, Lamport, Hanging Houghton, Maidwell, Draughton, Harrington, Kelmarsh, Braybrooke, Arthingworth, Clipston, Great Oxenden, East Farndon, Little Bowden (Northants.); terminating in a junction with the Rugby & Stamford Branch of the LNWR near Market Harborough.

67. Midland Railway. Extension from near Leicester via Bedford to Hitchin, and branches; and approaches to the intended station at Wellingborough. Charles Liddell, engineer. Deposited 30 Nov. 1852. Act 16 & 17 Vict. c. cviii (1853).

Main Line. From a junction with the Midland Railway near Wigston Station, through Wigston Magna, Newton Harcourt, Wistow, Burton Overy, Kibworth Beauchamp, Kibworth Harcourt, Church Langton, Tur Langton, East Langton, West Langton, Thorpe Langton, Great Bowden, Foxton, Market Harborough (Leics.); Little Bowden, Braybrooke, Desborough, Rushton, Barford, Glendon, Rothwell, Kettering, Broughton, Barton Seagrave, Pytchley, Burton Latimer, Isham, Finedon, Little Harrowden, Great Harrowden, Wellingborough, Irthlingborough, Irchester (Northants.); Wymington, Podington, Souldrop, Sharnbrook, Felmersham, Milton Ernest, Pavenham, Oakley, Clapham, Bromham, Biddenham, Bedford, Goldington, Elstow, Cardington, Eastcotts, Old Warden, Southill, Campton, Shefford, Shefford Hardwick, Meppershall, Henlow (Beds.); Holwell, Ickleford, Hitchin (Herts.); terminating in a junction with the Great Northern Railway near Hitchin Station.

Branches etc. From main line in Wellingborough, through Irchester; terminating in a junction with the Northampton & Peterborough branch of the London & North Western Railway at that company's Wellingborough Station in Irchester. And another branch connecting this branch with the main line, in Irthlingborough and Wellingborough (Northants.). And to make approach roads to an intended Midland Railway station at Wellingborough.

68. South Wales & Northamptonshire Junction Railway.

Liddell & Gordon, William Woodhouse, engineers. C.F. Cheffins, surveyor. Deposited 30 Nov. 1853. No Act.

From a junction with the Buckinghamshire Railway at Banbury Station, through Warkworth, Grimsbury (Northants.); Banbury, Neithrop, Broughton, North Newington, Bloxham, Tadmarton, Swalcliffe, Sibford Gower, Sibford Ferris, Hook Norton (Oxon.); Brailes, Whichford, Stourton, Sutton under Brailes, Cherington, Burmington, Tidmington (Wrwicks.); Todenham, Stretton on Foss, Blockley (Gloucs.); terminating in a junction with the Oxford, Worcester & Wolverhampton Railway at Blockley Station.

69. Stamford & Essendine Railway. Additional lands required for station and other purposes. William Hirst, surveyor. Deposited 29 Nov. 1856. Act 20 & 21 Vict. c. lxxxii (1857).

From a junction with the Midland Railway at Stamford Station in Stamford Baron (Northants.), to a junction with the Great Northern Railway at Essendine Station in Essendine (Rutland).

70. London & North Western Railway. Additional Works. Deposited 30 Nov. 1857. Act 21 & 22 Vict. c. cxxxi (1858).

Includes power to purchase land in the parishes of St Peter, Northampton (on Black Lyon Hill, adjoining the eastern approach to West Bridge, near Bridge Street Station), and Pitsford (Northants.) (on the Pitsford to Chapel Brampton road between the river Nene and a bridge in course of contruction over the company's railway), both on the Northampton to Market Harborough branch.

71. Northampton Gas Light Co. Additional Land. Deposited 28 Nov. 1857. Act 21 & 22 Vict. c. ciii (1858).

Additional land in Northampton All Saints, adjoining the existing gas works to the south, bounded on the south by the river Nene, with access on the west from Tanner Street and on the east by Weston Street.

72. Nene Valley Drainage & Navigation Improvements. John Fowler, engineer. Deposited 30 Nov. 1858. No Act.

Includes lands in Peterborough (Northants.); Fletton, Stanground (Cambs. & Hunts.), Whittlesey St Mary & St Andrew, Wisbech (Cambs.).

73. London & North Western Railway. Denton to Stalybridge [etc.]. Deposited 29 Nov. 1858. Act 22 & 23 Vict. c. cxiii (1859).

In Northants.: Additional lands and diversion of road in Lamport. With a plan showing proposed diversion of the road from Hanging Houghton to Cottesbrooke where it crosses the Northampton to Market Harborough line.

74. Weedon & Leamington Railway. William Jopling Nesham, engineer. R. & F. Hayward, surveyors. Deposited 30 Nov. 1858. No Act.

From a junction with the London & North Western Railway in Weedon Beck, through Dodford, Newnham, Daventry, Braunston (Northants.); Wolfhampcote, Grandborough, Napton on the Hill, Stockton, Southam, Long Itchington, Ufton, Radford Semele (Warwicks.); terminating there in a junction with the Rugby & Leamington Branch of the London & North Western Railway.

75. Northampton Waterworks. 1860. Act 24 & 25 Vict. c. xlvii (1861).

76. Stamford & Essendine Railway. Walter Marr Brydone, engineer. Deposited 30 Nov. 1861. No Act.

Main Line. From the existing S&E line at Stamford Station, through Stamford Baron (Northants.); Stamford, Newstead, Uffington (Lincs.); Pilsgate, Barnack, Southorpe, Bainton, Ufford, Wittering, Upton, Sutton, Castor, Ailsworth, Thornhaugh, Wansford (Northants.); Sibson cum Stibbington, Water Newton, Elton (Hunts.); terminating in Sutton (Northants.) at a junction with the Northampton and Peterborough branch of the London & North Western Railway near Sibson Station.

Branch. In Stamford Baron (Northants.), to connect the Stamford & Essendine Railway with the Syston and Peterborough line of the Midland Railway.

77. London & North Western Railway. Additional Powers. William Baker, engineer. Deposited 30 Nov. 1861. Act 24 & 25 Vict. c. ccviii.

In Northants.: Deviation of the Towcester to Cotton End Turnpike Road at Bridge Street Station, Northampton, and additional land in Hardingstone for sidings (detailed plan of station).

78. Daventry Railway. William Yates Freebody, engineer. Deposited 30 Nov. 1861. Act 25 & 26 Vict. c. lv (1862).

From a terminus at Daventry, through Newnham, Dodford, Norton, Brockhall, Weedon Beck (Northants.), terminating in a junction with the London & North Western Railway near Weedon Station.

79. Eastern Counties Railway. Railways between Wisbech and Peterborough. G.R. Stephenson, Robert Sinclair, engineers. C.F. Cheffins & Son, surveyors and lithographers. Deposited 30 Nov. 1861. No Act.

Railway No 1. From a junction with the Wisbech, St Ives & Cambridge Junction Railway of the ECR in Wisbech, through Guyhirn, Leverington, Murrow, Parson Drove, Whittlesey St Mary & St Andrew (Cambs.); Sutton St Edmund (Lincs.); terminating in Thorney (Cambs.).

Railway No 2. From a junction with Railway No 1 at its terminus as above, through Thorney Abbey, Whittlesey St Mary & St Andrew (Cambs.); Stanground (Cambs. & Hunts.); Eye, Newborough, Paston, Gunthorpe, Walton, Werrington, Peterborough (Northants.); terminating in Stanground by a junction with the Cambridge & Peterborough Railway of the ECR near the ECR station at Peterborough.

Also to alter the goods and coal branch of the Wisbech, St Ives & Cambridge Junction Railway, from its junction with the Wisbech branch of the East Anglian Railways up to its terminus on the bank of the river Nene, wholly in Wisbech.

80. Kettering & Thrapstone Railway. George B. Bruce, James Brunlees, engineers. Deposited 30 Nov. 1861. Act 25 & 26 Vict. c. clxxiii (1862).

From a junction with the Leicester & Hitchin line of the Midland Railway in Kettering, through Pytchley, Barton Seagrave, Burton Latimer, Cranford St John, Cranford St Andrew, Twywell, Woodford near Thrapston, Slipton, Great Addington, Islip, Thrapston (Northants.): terminating in a junction with the Northampton and Peterborough line of the London & North Western Railway.

81. Peterborough, Wisbeach & Sutton Railway. George B. Bruce, engineer. Deposited 29 Nov. 1862. Act 26 & 27 Vict. c. ccxxii (1863).

Railway No 1. From a terminus in Wisbech, through

Leverington, Tydd St Giles, Newton (Cambs.); Walsoken, West Walton, Walpole St Andrew, Walpole St Peter (Norfolk); Tydd St Mary, Long Sutton (Lincs.); terminating there in a junction with the Norwich & Spalding Railway near Sutton Bridge Station.

Railway No 2. From an end-on junction with Railway No 1 in Wisbech, through Guyhirn, Leverington, Murrow, Parson Drove, Whittlesey St Mary & St Andrew (Cambs.); Sutton St Edmund (Lincs.); terminating in Thorney Abbey near Thorney Gas & Waterworks.

Railway No 3. From an end-on junction with Railway No 2 in Thorney Abbey, through Whittlesey St Mary & St Andrew (Cambs.); Eye, Newborough, Paston, Gunthorpe, Walton, Werrington, Marholm, Peterborough (Northants.); terminating in a junction with the Syston & Peterborough branch of the Midland Railway.

Railway No 4. From a junction with Railway No 1 abutting on but to the south of the Norwich & Spalding Railway in Long Sutton terminating on the west side of the river Nene, all in Long Sutton (Lincs.).

82. London & North Western Railway. Additional Powers. William Baker, engineer. Deposited 27 Nov. 1862. Act 25 & 26 Vict. c. ccviii.

In Northants.: Diversion of the turnpike road from Towcester to Cotton End in Hardingstone at Bridge Street Station, with a plan of the station (similar to but less detailed than **77**).

83. Midland Railway. New Lines and Additional Powers. John S. Crossley, engineer. Deposited 29 Nov. 1862. Act 26 & 27 Vict. c. clxxxiii (1863).

In Northants.: (a) Siding on the Northampton & Peterborough branch of the London & North Western Railway in Hardingstone (Northants.), to the east of London Road, Northampton; (b) Additional land to the east of the Midland Railway's Wellingborough Station (simple plan of station); (c) Additional land to the west of the Midland Railway's Crescent Station at Peterborough (simple plans of Midland and Great Northern stations); (d) Additional land in Isham at Finedon Station on the Leicester to Hitchin line (simple plan of station).

84. Northampton & Banbury Junction Railway. John Collister, engineer. Deposited 29 Nov. 1862. Act 26 & 27 Vict. c. ccxx (1863).

From a junction with the London & North Western Railway near Blisworth Station, through Blisworth, Gayton, Tiffield, Caldecote, Towcester, Easton Neston, Bradden, Greens Norton, Abthorpe, Slapton, Wappenham, Weedon Lois, Helmdon, Astwell with Falcutt, Stuchbury, Greatworth, Farthinghoe (Northants.); terminating at Greatworth in a junction with the Banbury Extension of the Buckinghamshire Railway near Cockley Brake.

85. East & West Junction Railway. Blisworth to Worcester, with branches. No engineer named. Deposited 29 Nov. 1862. No Act.

Railway No 1. From a junction with the London & North Western Railway near Blisworth Station, through Blisworth, Gayton, Tiffield, Towcester, Greens Norton, Bradden, Blakesley, Adstone, Canons Ashby, Moreton Pinkney, Eydon, Woodford cum Membris, Byfield, Aston le Walls, Boddington (Northants.); Claydon (Oxon.); Wormleighton, Fenny Compton, Burton Dassett, Chadshunt, Kineton, Butlers Marston, Combrook, Wellesbourne Hastings, Pillerton Hersey, Eatington (Warwicks.); Alderminster (Worcs.); Preston on Stour (Gloucs.); Atherstone upon Stour (Warwicks.); Clifford Chambers (Gloucs.); Old Stratford & Drayton (Warwicks.); terminating in a junction with the West Midland Railway near the Stratford on Avon Station of that railway.

Railway No 2. From a junction with Railway No 1 near Fenny Compton Station on that railway, terminating in a junction with the Great Western Railway near the Fenny Compton Station of that railway, all in Fenny Compton (Warwicks.).

Railway No 3. From a junction with Railway No 1 in Old Stratford & Drayton (Warwicks.), through Shottery, Luddington, Binton, Temple Grafton, Bidford (Warwicks.); terminating at Bidford.

Railway No 4. From the termination of Railway No 3, through Bidford, Salford Priors (Warwicks.); Rous Lench, Abbots Morton, Inkberrow, Dormston, Flyford Flavel, Grafton Flyford, North Piddle, Upton Snodsbury, Broughton Hackett, Churchill, Bredicot, Tibberton, Spetchley, North Claines (Worcs.); terminating in Worcester in a junction with the West Midland Railway near the Worcester Station of that railway.

Railway No 5. From the termination of Railway No 3, through Bidford, Wixford, Arrow, Oversley, Alcester, Coughton, Sambourne, Studley, Ipsley (Warwicks.); Redditch, Tardebigge (Worcs.); terminating at Redditch in a junction with the Redditch Railway near the Redditch Station of that railway.

Railway No 6. From the termination of Railway No 3, through Bidford, Abbots Salford, Salford Priors (Warwicks.); Cleeve Prior, Harvington, Norton juxta Kempsey, Lenchwick, Cropthorne, Great & Little Hampton, Evesham (Worcs.); terminating at Evesham in a junction with the authorised line of the Ashchurch & Evesham Railway, near the Evesham Station of the West Midland Railway.

Railway No 7. From a junction with Railway No 1 in Canons Ashby, through Moreton Pinkney, Culworth, Thorpe Mandeville, Chacomb, Middleton Cheney, Grimsbury, Warkworth (Northants.); Wardington, Hanwell, Neithrop, Hardwick (Oxon.); terminating in a junction with the Great Western Railway near the Banbury Station of that railway.

Railway No 8. From a junction with Railway No 7 in Grimsbury, terminating in a junction with the Buckinghamshire Railways at the Banbury Station of that railway in Warkworth (Northants.).

86. Market Harborough & Melton Mowbray Railway. Nixon & Dennis, engineers. Alexander & Littlewood, surveyors and lithographers. Deposited 29 Nov. 1862. No Act.

From a junction with the Rugby & Stamford line of the London & North Western Railway in Ashley near Medbourne Bridge Station (Northants.), through Weston by Welland, Ashley (Northants.); Medbourne, Welham, Slawston, Blaston St Michael, Blaston St Giles, Hallaton, Horninghold, Allexton, East Norton, Tugby, Skeffington, Loddington, Withcote, Launde, Tilton, Halstead, Marefield, Whatborough, Owston & Newbold, Lowesby, Cold Newton, Twyford, South Croxton, Thorpe Satchville, Burrough on the Hill, Ashby, Floville, Gaddesby, Great Dalby, Kirby Bellars, Burton Lazars, Sysonby, Melton Mowbray, Freeby, Welby, Asfordby, Saxelby (Leics.); terminating at Melton Mowbray in a junction with the Syston & Peterborough branch of the Midland Railway west of the Melton Mowbray Station on that branch.

87. Kettering & Thrapstone Railway. Huntingdon Extension. George C. Bruce, James Brunlees, engineers. Deposited 29 Nov. 1862. No Act.

Railway No 1. From the authorised line of the Kettering &

Thrapstone Railway in Islip, through Woodford near Thrapston, Islip, Thrapston, Denford, Ringstead, Raunds, Stanwick (Northants.); Keyston, Hargrave, Covington, Catworth (Hunts.); Hargrave (Northants.), Tilbrook, Kimbolton, Stow, Easton, Spaldwick, Grafham, Buckden, Brampton, Offord Cluny, Godmanchester, Huntingdon (Hunts.); Dean (Beds.); terminating in a junction with the St Ives & Huntingdon branch of the Great Eastern Railway in Huntingdon.

Railway No 2. From a junction with Railway No 1 to a junction with the Great Northern Railway near that company's Huntingdon Station, all in Huntingdon.

88. Market Harborough & East Norton Railway.
John Addison, engineer. Deposited 30 Nov. 1863. No Act.

From a junction with the Rugby & Stamford line of the London & North Western Railway near Medbourne Bridge Station, through Weston by Welland, Ashley (Northants.); Medbourne, Welham, Slawston, Blaston, Hallaton, Horninghold, Allexton, East Norton (Leics.); terminating in East Norton.

89. Great Eastern Northern Junction Railway.
Messrs Hawkshaw, Fowler & Stephenson, engineers-in-chief. Messrs Fraser & Stanton, acting engineers. Deposited 30 Nov. 1863. No Act.

Railway No 1. From a junction with the authorised line of the West Riding & Grimsby Railway in Owston (Yorks.); through Doncaster, Barnby upon Don, Kirk Sandall, Armthorpe, Cantley (Yorks.); Finningley (Notts.); Misson (Notts.); Haxey (Lincs.); Misterton, Walkeringham, Beckingham, Saundby (Notts.); Gainsborough, Corringham, Heapham, Upton, Willingham, Stowe, Thorpe in the Fallows, Scampton, Broxholme, North Carlton, South Carlton, Burton by Lincoln, Lincoln, Skellingthorpe, Boultham, Bracebridge, Waddington, Harmston, Coleby, Boothby Graffoe, Navenby, Temple Bruer with Temple High Grange, Ashby de la Launde, Brauncewell, Roxholm, Leasingham, Holdingham, New Sleaford, Old Sleaford, Silk Willoughby, Burton Pedwardine, Scredington, Spanby, Threekingham, Horbling, Billingborough, Sempringham, Aslackby, Dowsby, Rippingale, Dunsby, Hacconby, Morton, Bourne, Thurlby, Baston, Langtoft, Market Deeping, Deeping St James (Lincs.); Maxey, Northborough, Glinton, Peakirk, Werrington, Paston, Peterborough (Northants.); Stanground (Cambs. & Hunts.); Farcet, Ramsey, Bury, Wistow, Warboys, Pidley cum Fenton, Somersham, Colne, Bluntisham cum Earith (Hunts.); Over, Willingham, Long Stanton All Saints (Cambs.); terminating in a junction with the St Ives & Cambridge line of the Great Eastern Railway in Long Stanton All Saints, Over and Rampton (Cambs.).

Railway No 2. From a junction with Railway No 1 in Blaxton (Yorks.); through Finningley (Notts.); Cantley, Carr House and Elmfield, Doncaster, Balby with Hexthorpe (Yorks.); terminating in a junction with the South Yorkshire Railway in Balby with Hexthorpe.

Railway No 3. From a junction with Railway No 2 in Doncaster; through Balby with Hexthorpe; terminating in a junction with the Great Northern Railway and South Yorkshire Railway in Doncaster (Yorks.).

Railway No 4. From a junction with Railway No 1 in Barnby; through Doncaster, Owston, Burghwallis (Yorks.); terminating in a junction in Owston and Burghwallis with the Lancashire & Yorkshire Railway.

Railway No 5. From a junction with Railway No 1 in Doncaster (Yorks.); through Kirk Sandall, Barnby Dun; terminating in a junction with the authorised line of the South Yorkshire Railway (Thorne Branch) in Barnby Dun.

Railway No 6. From a junction with Railway No 1 in Beckingham (Notts.); through Saundby, Bole (Notts.); terminating in a junction with the Manchester, Sheffield & Lincolnshire Railway in Bole.

Railway No 7. From a junction with Railway No 1 in Gainsborough (Lincs.); terminating in a junction with the Manchester, Sheffield & Lincolnshire Railway in Gainsborough.

Railway No 8. From a junction with Railway No 1 to a junction with the Great Northern Railway, all in Skellingthorpe.

Railway No 9. From a junction with Railway No 1 in Skellingthorpe; through Lincoln, Boultham, Skellingthorpe (Lincs.); terminating in a junction with the Midland Railway in Boultham.

Railway No 10. From a junction with Railway No 1 in Bracebridge (Lincs.); through Lincoln, Canwick, Monk's Liberty, Greetwell (Lincs.); terminating in a junction with the Manchester, Sheffield & Lincolnshire Railway in Greetwell.

Railway No 11. From a junction with Railway No 10 in High Street, Lincoln; through Lincoln (Lincs.); terminating in a junction with the Manchester, Sheffield & Lincolnshire Railway in St Peter at Gowts, Lincoln.

Railway No 12. From a junction with Railway No 1 between Sleaford Lodge and Sleaford; terminating in a junction with the Boston, Sleaford & Midland Counties Railway at a level crossing on the road from Folkingham to Sleaford, all in Old Sleaford.

Railway No 13. From a junction with Railway No 1 at Tunnel Bank; terminating in a junction with the authorised line of the Spalding & Bourn Railway, all in Bourn (Lincs.).

Railway No 14. From a junction with Railway No 1 near the old river Nene near Stanground Sluice; terminating in a junction with the Peterborough & Ely line of the Great Eastern Railway to the east of that company's Peterborough Station; all in Stanground (Cambs. & Hunts.).

Railway No 15. From a junction with Railway No 1 at a road leading from Lodge House to Somersham; terminating in a junction with the St Ives & March line of the Great Eastern Railway near the 77th milepost; all in Somersham (Hunts).

90. Stamford & Essendine Railway. Sibson Extension.
Walter Marr Brydone, engineer. Deposited 28 Nov. 1863. Act 27 & 28 Vict. c. ccxx (1864).

From a junction with the S&E near Stamford Station in Stamford Baron (Northants.), through Stamford, Newstead, Uffington (Lincs.); Pilsgate, Barnack, Southorpe, Ufford, Bainton, Wittering, Upton, Sutton, Castor, Ailsworth, Thornhaugh, Wansford (Northants.); Sibson cum Stibbington (Hunts.); terminating in Sutton (Northants.) in a junction with the Northampton & Peterborough branch of the London & North Western Railway near Sibson Station.

91. Market Harborough & Melton Mowbray Railway.
Joseph Cubitt, John Wright, engineers. Deposited 30 Nov. 1863. No Act.

From a junction with the Rugby & Stamford line of the London & North Western Railway near Medbourne Bridge Station in Ashley, through Weston by Welland (Northants.); Medbourne, Welham, Slawston, Blaston St Michael, Blaston St Giles, Hallaton, Horninghold, Allexton, East Norton, Tugby, Skeffington, Loddington, Withcote, Launde, Tilton, Halstead, Marefield, Whatborough, Owston & Newbold, Lowesby, Cold Newton, Twyford, South Croxton, Thorpe Satchville, Burrough on the Hill, Ashby Folville, Gaddesby, Great Dalby, Little Dalby, Somerby, Kirby Bellars, Burton Lazars, Sysonby,

Melton Mowbray, Ab Kettleby, Freeby, Welby, Asfordby, Saxelby (Leics.), terminating at Melton Mowbray in a junction with the Syston and Peterborough branch of the Midland Railway near Melton Mowbray Station.

92. Blockley & Banbury Railway. Charles Liddell, John Collister, engineers. Charles R. Cheffins, surveyor, London. Deposited 30 Nov. 1863. No Act.

Railway No 1. From a junction with the Great Western Railway in Blockley (Gloucs.), through Tidmington, Stretton on Fosse, Long Compton, Burmington (Gloucs.); Cherrington (Warwicks.); Whichford (Gloucs.); Brailes, Sutton under Brailes, Barcheston, Little Wolford, Great Wolford, Stourton (Warwicks.); Todenham, Lower Lemington (Gloucs.); Warkworth, Middleton Cheney, Marston St Lawrence, Grimsbury, King's Sutton (Northants.); Swalcliffe, Banbury, Neithrop, Tadmarton, Sibford, Sibford Gower, Sibford Ferris, Hook Norton, Bloxham, Bodicote, Broughton, West Shutford, Epwell, Wigginton, Milcombe, Drayton, Adderbury (Oxon.); and terminating in Middleton Cheney (Northants.) by a junction with the Banbury Extension of the Buckinghamshire Railway.

Railway No 2. From a junction with the Great Western Railway near Banbury Station, terminating in a junction with Railway No 1, all in Warkworth (Northants.)

93. Daventry Railway. Extensions to Southam and Leamington. G.W. Hemans, engineer. Deposited 30 Nov. 1863. No Act.

Railway No 1. From the line of the Daventry Railway authorised by Daventry Railway Act 1862, in Daventry, through Newnham, Braunston, Welton (Northants.); Wolfhampcote, Nethercote, Willoughby, Grandborough, Upper Shuckburgh, Lower Shuckburgh, Caldecote, Napton on the Hill, Long Itchington and Southam (Warwicks.), terminating there.

Railway No 2. From the terminus of Railway No 1, through Southam, Ufton, Long Itchington, Offchurch, Radford Semele (Warwicks.), terminating there in a junction with the Rugby & Leamington Railway of the London & North Western Railway.

94. East & West Junction Railway. James B. Burke, engineer. Deposited 30 Nov. 1863. Act 27 & 28 Vict. c. lxxvi (1864).

Railway No 1. From a junction with the authorised line of the Northampton & Banbury Junction Railway in Towcester, through Greens Norton, Bradden, Woodend, Blakesley, Adstone, Moreton Pinkney, Canons Ashby, Eydon, Woodford cum Membris, Hinton in the Hedges, Plumpton, Byfield, Aston le Walls, Appletree, Boddington (Northants.); Claydon, Cropredy (Oxon.); Wormleighton, Farnborough, Fenny Compton, Burton Dassett, Gaydon, Chadshunt, Kineton, Butlers Marston, Combrook, Wellesbourne Hastings, Walton, Pillerton Hersey, Atherstone on Stour, Loxley, Eatington, Old Stratford & Drayton (Warwicks.); Alderminster (Worcs.); Preston on Stour, Clifford Chambers (Gloucs.), terminating in a junction with the Great Western Railway (Stratford & Honeybourne Branch) in Old Stratford & Drayton near the GWR Stratford on Avon Station.

Railway No 2. From a junction with Railway No 1 in Old Stratford & Drayton, through Luddington, Binton, Temple Grafton, Bidford, Salford Priors (Warwicks.), terminating there in a junction with the authorised line of the Evesham & Redditch Railway.

94a. Chipping Norton & Banbury Railway. John Fowler and Edward Wilson, engineers. Deposited 30 Nov. 1863. No Act.

From a junction with the Chipping Norton branch of the Great Western Railway at Chipping Norton Station, through Chipping Norton, Over Norton, Heythrop, Great Rollright, Little Rollright, Swerford, Little Tew, Wiggington, South Newington, Bloxham, Milcomb, Adderbury, Barford St John, Milton, Bodicote (Oxon.); King's Sutton (Northants.), terminating there in a junction with the Oxford & Birmingham line of the GWR near Twyford Mill.

95. Northampton & Banbury Railway. Deviations &c. John Collister, engineer. Alexander & Littlewood, surveyors. Deposited 30 Nov. 1864. Act 28 & 29 Vict. c. ccclxi (1865).

Deviation Railway. From a junction with the Great Western Railway north of that company's Banbury Station in Warkworth (Northants.); through Grimsbury, Middleton Cheney, Marston St Lawrence, Chacomb, Thorpe Mandeville, Culworth, Sulgrave, Weedon Lois, Helmdon, Wappenham, Astwell with Falcutt (Northants.); Banbury, Neithrop, Hardwick, Cropredy, Wardington (Oxon.); terminating in Wappenham in a junction with the authorised line of the Northampton & Banbury Junction Railway, as per plan deposited Nov. 1862 [**84**].

Branch Railway No 1. From a junction with the Banbury Extension of the Buckinghamshire Railway at that company's Banbury Station (plan of station), terminating with Railway No 1 about 680 yards north of the GWR Banbury Station; all in Warkworth and Grimsbury (Northants.).

Branch Railway No 2. From a junction with the authorised line of the Northampton & Banbury Junction Railway in Gayton; terminating in Blisworth in a junction with the Northampton and Peterborough line of the London & North Western Railway near Gayton Wharf.

96. Banbury Water. Plans and sections of the works for the supply of water to the town of Banbury and other places in the counties of Oxford and Northampton. T. Hawksley, engineer. Deposited 30 Nov. 1864. Act 28 & 29 Vict. c. xvi.

Works in Warkworth and Grimsbury (Northants.); Banbury, Neithrop (Oxon.); including waterworks near Grimsbury Mills, reservoir near Easington Farm, and pipes through streets between these points.

97. Lancashire & Yorkshire and Great Eastern Junction Railway. John Hawkshaw, Robert Sinclair, engineers. Charles R. Cheffins, surveyor, London. Deposited 30 Nov. 1864. No Act.

Railway No 1. From a junction with the St Ives & Cambridge branch of the Great Eastern railway near Long Stanton Station in Long Stanton All Saints; through Over, Willingham (Cambs.); Bluntisham cum Earith, Colne, Somersham, Pidley cum Fenton, Warboys, Wistow, Bury, Ramsey, Farcet (Hunts.), Stanground (Cambs. & Hunts.); Peterborough, Paston, Werrington, Glinton, Peakirk, Northborough, Maxey (Northants.); Market Deeping, Deeping St James, Langtoft, Baston, Thurlby, Bourn, Morton, Haconby, Dunsby, Rippingale, Dowsby, Aslackby, Sempringham, Billingborough, Horbling, Threekingham with Stowe, Spanby, Scredington, Silk Willoughby, Old Sleaford, Burton Pedwardine, Kirkby Laythorpe, New Sleaford, Holdingham, Leasingham, Ruskington, Bloxholm, Digby, Roulston, Scopwick, Kirkby Green, Blankney, Metheringham, Dunston, Nocton, Potter Hanworth, Branston, Heighington, Washingborough, Canwick, Lincoln, Burton, South Carlton, North Carlton, Broxholme, Scampton, Stowe, Thorpe in the Fallows, Willingham, Upton, Heapham, Springthorpe, Great Gorringham, Pilham, Blyton, Laughton, Owston, Haxey,

Epworth, Wroot (Lincs.); Hatfield, Fishlake, Armthorpe, Barnby Dun (Yorks.); terminating in a junction with the Askern branch of the Lancashire & Yorkshire Railway in Burghwallis.

Railway No 2. From a junction with Railway No 1 to a junction with the St Ives & March Railway belonging to the GER, all in Somersham (Hunts.).

Railway No 3. From a junction with Railway No 1 to a junction with the Ramsey Railway near Ramsey Station, all in Ramsey (Hunts.)

Railway No 4. From a junction with Railway No 1 to a junction with the Peterborough & Ely branch of the GER, all in Stanground (Cambs. & Hunts.).

Railway No 5. From a junction with Railway No 1 to a junction with the authorised line of the Spalding & Bourn Railway, all in Bourn (Lincs.).

Railway No 6. From a junction with Railway No 1 to a junction with the Boston, Sleaford & Midland Counties Railway at Sleaford Station, all in Old Sleaford (Lincs.).

Railway No 7. From a junction with Railway No 1 to a junction with the Manchester, Sheffield & Lincolnshire Railway, all in Monk's Liberty (Lincs.).

Railway No 8. From a junction with Railway No 1 in Pilham (Lincs.), terminating in Blyton (Lincs.) in a junction with the MS&LR near Blyton Station.

Railway No 9. From a juncton with Railway No 1 to a junction with the MS&LR near Blyton Station, all in Blyton (Lincs.).

Railway No 10. From a junction with Railway No 1 to a junction with the authorised line of the South Yorkshire Railway, all in Barnby Dun (Yorks.).

Railway No 11. From a junction with Railway No 1 to a junction with the West Riding & Grimsby Railway, all in Barnby Dun (Yorks.).

Railway No 12. From a junction with Railway No 1 in the township of Thorpe in Balne in Barnby Dun (Yorks.), terminating in a junction with the authorised York and Doncaster branch of the North Eastern Railway in Burghwallis (Yorks).

Which railways will pass through Long Stanton All Saints, Long Stanton St Michael, Swavesey, Rampton, Willingham, Over, Whittlesey St Mary & St Andrew (Cambs.); Stanground (Cambs. & Hunts.); Bluntisham cum Earith, Colne, Somersham, Pidley cum Fenton, Warboys, Wistow, Bury, Ramsey, Upwood, Holme, Farcet, Fletton (Hunts.); Peterborough, Walton, Paston, Gunthorpe, Werrington, Marholm, Newborough, Borough Fen, Glinton, Helpston, Peakirk, Etton, Northborough, Deeping Gate, Maxey (Northants.); Deeping St James, Market Deeping, Langtoft, Baston, Thurlby, Northorpe, Bourne, Morton, Hacconby, Dunsby, Rippingale, Kirkby Underwood, Dowsby, Aslackby, Sempringham, Pointon, Birthorpe, Laughton, Folkingham, Billingborough, Horbling, Swaton, Threekingham, Spanby, Osbournby, Aswarby, Scredington, Burton Pedwardine, Silk Willoughby, Quarrington, Sleaford, Old Sleaford, New Sleaford, Kirkby La Thorpe, Holdingham, Evedon, Leasingham, Roxholm, Ruskington, Anwick, Cranwell, Dorrington, Broxholme, Ashby, Digby, Rowlston, Kirkby Green, Scopwick, Blankney, Metheringham, Dunston, Nocton, Potter Hanworth, Heighington, Washingborough, Cherry Willingham, Greetwell, Burton by Lincoln, South Carlton, North Carlton, Broxholme, Scampton, Thorpe in the Fallows, Brattleby, Stow, Sturton, Coates, Aisthorpe, Willingham, Upton, Kexby, Heapham, Lea, Springthorpe, Corringham, Somerby, Gainsborough, Pilham, Blyton, Scotter, Loughton, Wildsworth, Owston Ferry, Haxey, Epworth, Wroot, Lincoln (Lincs.); Hatfield, Stainforth, Fishlake, Sykehouse, Armthorpe, Kirk Sandall, Barnby Dun, Thorpe in Balne, Kirk Bramwith, Burghwallis, Owston, Skellow, Campsall, Moss and Askern (Yorks.).

98. Bourton, Chipping Norton & Banbury Railway. John Fowler, Edward Wilson, engineers. Deposited 30 Nov. 1864. No Act.

Railway No 1. From a junction with the Bourton on the Water Railway in Bledington (Gloucs.), near Chipping Norton Junction Station; terminating in a junction with the West Midland line of the Great Western Railway in Kingham (Oxon.).

Railway No 2. From the commencement of Railway No 1; terminating in a junction with the Chipping Norton branch of the GWR in Churchill (Oxon.).

Railway No 3. From the termination of Railway No 1 to the termination of Railway No 2.

Railway No 4. From the terminations of Railways No 2 and 3; through Churchill (Oxon.); terminating at Chipping Norton (Oxon.) in a junction with the Chipping Norton branch of the GWR.

Railway No 5. From the termination of Railway No 4; through Over Norton, Great Rollright, Hook Norton, Swalcliffe, Tadmarton, Bloxham, Banbury, Neithrop, Bodicote (Oxon.); terminating in Bodicote.

Railway No 6. From an end-on junction with Railway No 5; terminating in Warkworth (Northants.), near the Banbury Station of the GWR in a junction with the GWR.

Railway No 7. From the termination of Railway No 5; terminating in Warkworth in a junction with the Buckinghamshire Railway of the London & North Western Railway (good plan of both GWR and LNWR Banbury stations).

99. Bedford, Northampton & Weedon Railway. Charles Liddell, engineer. Charles R. Cheffins, surveyor. Deposited 30 Nov. 1864. Act 28 & 29 Vict. c. ccclv (1865).

Railway No 1. From a junction with the Leicester and Hitchin branch of the Midland Railway in Bromham, through Stevington, Turvey (Beds.); Newton Blossomville, Clifton Reynes, Olney, Weston Underwood, Ravenstone (Bucks.); Yardley Hastings, Horton, Piddington, Hackleton, Little Houghton, Great Houghton, Hardingstone, Northampton (St Giles, All Saints); terminating in a garden belonging to the Master and Co-Brethren of St John's Hospital (enlarged plan of built-up area).

Railway No 2. From a junction with Railway No 1 to a west-facing junction with the Blisworth and Peterborough branch of the London & North Western Railway, all in Hardingstone (Northants.).

Railway No 3. From a junction with Railway No 1 to an east-facing junction with the Blisworth and Peterborough branch of the London & North Western Railway, through Northampton (St Giles), terminating in Hardingstone (Northants.).

Railway No 4. From an end-on junction with the terminus of Railway No 1, in All Saints, through Duston, to a junction with the Northampton and Market Harborough branch of the LNWR near West Bridge in St Peter, Northampton.

Railway No 5. From a junction with Railway No 4 in Northampton All Saints on a piece of land belonging to the Northampton Gas Light Co. on Mill Lane, through Northapton St Peter and Duston, to a junction with the Northampton and Market Harborough branch of the LNWR near where the branch from the gas company's yards forms a junction with the LNWR branch, in Hardingstone (Northants.).

Railway No 6. From a junction with Railway No 4 at the

same point as the start of Railway No 5, in Northampton All Saints, through Northampton St Peter, Duston, Upton, Kislingbury, Bugbrooke, Upper Heyford, Nether Heyford, Flore, Stowe Nine Churches, to a junction with the authorised Weedon & Daventry Railway (as shown in a plan deposited in Nov. 1861 [**78**]) in Dodford (Northants.).

Railway No 7. From a junction with Railway No 6 in Upton near Upton Mill, through Duston, terminating in Hardingstone in a junction with the Blisworth and Peterborough branch of the LNWR near milepost 66.

Railway No 8. From a junction with Railway No 6 in Dodford, through Weedon Beck, terminating in a junction with the main line of the LNWR in Dodford, near Weedon Station.

Railway No 9. From a junction with Railway No 6 in Dodford, terminating in the same parish in a junction with the main line of the LNWR.

100. Peterborough, Wisbech & Sutton Railway. Extensions and Deviations. George B. Bruce, engineer. Charles R. Cheffins, surveyor. Deposited 30 Nov. 1864. Act 28 & 29 Vict. c. cccxl (1865).

Railway No 1. From a junction with the Great Eastern Railway at a junction with the authorised line of the Peterborough, Wisbech & Sutton Railway (as per plan deposited Nov. 1862 [**81**]) and near that company's Wisbech Station, through Wisbech, Elm (Cambs.); Emneth (Norfolk); Outwell (Cambs.); and Upwell (Cambs. and Norfolk), terminating there at the Wisbech to Ely turnpike road.

Railway No 2. From a junction with the authorised line of the Peterborough, Wisbech & Sutton Railway in Eye, through Paston, Werrington, Peakirk, Glinton, Newborough, Borough Fen (Northants.); Crowland, Deeping St James, Deeping Fen, Cowbit, Deeping St Nicholas (Lincs.); terminating in Crowland near Crowland Abbey.

Deviation near Sutton Bridge. From a junction with the authorised line (as per deposited plan of 1862) to a new junction with the Norwich and Spalding Railway near that company's Sutton Bridge Station, all in Long Sutton (Lincs.).

101. Chipping Norton, Banbury & East & West Junction Railway. James B. Burke, engineer. Deposited 30 Nov. 1864. No Act.

Railway No 1. From a junction with the authorised line of the East & West Junction Railway (Act, 1864; plan **94**) in Canons Ashby; through Moreton Pinkney, Culworth, Thorpe Mandeville (Northants.); Wardington (Oxon.); Chalcombe (Northants.); Neithrop, Hardwick (Oxon.); Warkworth, Grimsbury (Northants.); terminating there in a junction with the GWR near milepost 87.

Railway No 2. From a junction with the GWR near that company's Banbury Station; through Warkworth, Grimsbury (Northants.); Bodicote, Adderbury, Neithrop, Bloxham, Tadmarton, Swalcliffe, Hook Norton (Oxon.); Whichford (Warwicks.); Great Rollright, Over Norton, Salford, Chipping Norton (Oxon.); terminating in a junction with the Chipping Norton branch of the GWR near that company's Chipping Norton Station.

Railway No 3. From a junction with Railway No 1 in Grimsbury, through Neithrop, to a junction with Railway No 2 near the GWR Banbury Station.

102. Northampton & Banbury Junction Railway. Extensions. John Collister, engineer. Alexander & Littlewood, surveyors. Deposited 30 Nov. 1864. Act 28 & 29 Vict. c. ccclxii.

Railway No 1. From a junction with the Chipping Norton branch of the GWR near the station there, through Over Norton, Salford, Great Rollright, Little Rollright, Heythrop, Swerford, Hook Norton, Wigginton, South Newington, Milcomb; terminating at Bloxham (Oxon.).

Railway No 2. From an end-on junction with Railway No 1 at Bloxham, through Bodicote, Adderbury, Broughton, Neithrop, Banbury (Oxon.); Warkworth, Grimsbury (Northants.); terminating in Warkworth in a junction with the GWR near that company's Banbury Station (plans of both Banbury GWR and Buckinghamshire Railway stations).

Railway No 3. From a junction with Railway No 2 in Bodicote (Oxon.); through Middleton Cheney, Warkworth (Northants.); Banbury (Oxon.); Grimsbury (Northants.); Bodicote and Adderbury (Oxon.); terminating in Middleton Cheney (Northants.) in a junction with the Banbury Extension Line of the Buckinghamshire Railway near milepost 77.

Railway No 4. From a point in Warkworth near Banbury GWR Station, through Banbury, Grimsbury, Middleton Cheney (Northants.); Neithrop, Banbury, Bodicote (Oxon.); terminating in Warkworth in a junction with Railway No 3.

Railway No 5. From a junction with Railway No 4 in Warkworth, through Banbury, Grimsbury (Northants.); Neithrop, Banbury, Bodicote, Adderbury (Oxon.); terminating in a junction with Railway No 2 in Bodicote.

Railway No 6. From a junction with the Great Western Railway in Blockley (Gloucs.); through Tidmington (Warwicks.); Lower Lemington, Todenham, Stretton on Fosse (Gloucs.); Burmington, Great Wolford, Long Compton, Stourton, Cherington, Whichford, Brailes (Warwicks.); Hook Norton, Swalcliffe, Tadmarton, Milcombe, Bloxham (Oxon.); terminating in an end-on junction with Railway No 1.

103. Bedford, Northampton & Leamington Railway. Sir Charles Fox & Son, engineers. Deposited 30 Nov. 1864. No Act.

Railway No 1. From Bedford, making a level crossing with the Midland Railway on the south side of Bedford Station, through Biddenham, Bromham, Kempston, Stagsden, Turvey (Beds.); Newton Blossomville, Clifton Reynes, Olney (Bucks.); terminating at Olney.

Railway No 2. From an end-on junction with Railway No 1 at Olney, through Weston Underwood, Ravenstone, Stoke Goldington, Lavendon (Bucks.); Cogenhoe, Horton, Piddington, Preston Deanery, Wootton, Great Houghton, Hardingstone, Northampton (All Saints); terminating in All Saints on the west side of Cow Lane. Enlarged details of lines within Northampton.

Railway No 2a. From a junction with Railway No 2, to a junction with the Northampton and Peterborough railway, all in Hardingstone.

Railway No 3. From a junction with Railway No 2, terminating in a junction with Railway No 5, all in Northampton All Saints.

Railway No 4. From an end-on junction with Railway No 2, and at the commencement of Railway No 5, in All Saints, through St Giles, terminating on the south side of St Giles Square, all in Northampton.

Railway No 5. From an end-on junction with Railway No 2, and at the commencement of Railway No 4, in All Saints, through Hardingstone, Duston, Upton, Kislingbury, Harpole, Bugbrooke, Nether Heyford, Upper Heyford, Flore, Stowe Nine Churches, Dodford (Northants.); terminating in a junction with the authorised Daventry Railway now in course of construction.

Railway No 6. From a junction with the Daventry Railway

in Daventry, through Braunston (Northants.); Wolfhampcote (Warwicks.); Grandborough, Caldecote, Napton on the Hill, Stockton, Southam (Warwicks.); terminating at Southam near the union workhouse.

Railway No 7. From an end-on junction with Railway No 6 at Southam, through Long Itchington, Ufton, Offchurch, Radford Semele, Whitnash, Chesterton & Kingston, Harbury, Leamington Priors (Warwicks.); terminating at Leamington.

Railway No 8. From an end-on junction with Railway No 7, terminating in a junction with the railway from Leamington to Rugby, all in Leamington Priors (Warwicks.).

Railway No 9. From an end-on junction with Railway No 7, terminating in a junction with the Great Western Railway from Oxford to Leamington, all in Leamington Priors (Warwicks.).

Railway No 10. From a junction with Railway No 1 to a junction with the Midland Railway Bedford and Wellingborough line near the MR Bedford Station, in Bedford.

104. East & West Junction Railway. Extension. James B. Burke, engineer. Deposited 30 Nov. 1864. No Act.

Railway No 1. From a junction in Blisworth or Gayton with the authorised line of the Northampton & Banbury Railway; through Rothersthorpe, Kislingbury, Upton, Duston, Northampton, Abington, Hardingstone, Great Houghton, Little Houghton, Brafield on the Green, Denton, Castle Ashby, Yardley Hastings (Northants.); Olney, Warrington, Clifton Reynes, Lavendon, Cold Brayfield (Bucks.); Turvey, Stevington, Bromham, Biddenham (Beds.); terminating in Bedford in a junction with the Midland Railway Leicester and Hitchin line, near the MR Bedford Station.

Railways No 2 and 3. Short lines, wholly within Gayton, forming with Railway No 1 a triangular junction with the Northampton & Banbury Railway.

Railway No 4. From a junction with Railway No 1 in Northampton St Peter, terminating in Duston (Northants.) in a junction with the Northampton and Market Harborough branch of the London & North Western Railway south of the LNWR Castle Hill Station.

Railway No 5. From a junction with Railway No 1, to a junction with the Northampton and Peterborough branch of the LNWR, all in Great Houghton (Northants.).

105. Chipping Norton, Banbury, East & West Junction Railway. James B. Burke, engineer. Deposited 30 March (*sic*) 1865. No Act.

Railway No 1. From a junction with the authorised line of the East & West Junction Railway (Act, 1864; plan **94**) in Canons Ashby; through Moreton Pinkney, Culworth, Thorpe Mandeville (Northants.); Wardington (Oxon.); Chalcombe (Northants.); Neithrop, Hardwick (Oxon.); Warkworth, Grimsbury (Northants.); terminating in a junction with the Great Western Railway near milepost 87 in Grimsbury (Northants.) (including plans of GWR and LNWR stations at Banbury).

Railway No 2. From a junction with the GWR near Banbury Station in Grimsbury (Northants.); through Adderbury, Bodicote, Neithrop, Bloxham, Tadmarton, Swalcliffe, Hook Norton (Oxon.); Whichford (Warwicks.); Great Rollright, Over Norton, Salford, Chipping Norton (Oxon.); terminating in a junction with the Chipping Norton branch of the GWR near Chipping Norton Station.

Railway No 3. From an end-on junction with Railway No 1 to a junction with Railway No 2, in Neithrop and Grimsbury.

106. Bedford & Northampton Railway. Deviations &c. Charles Liddell, engineer. Deposited 30 Nov. 1865. Act 29 & 30 Vict. c. cclx (1866).

Railway No 1. From a junction with the railway authorised by the Bedford & Northampton Railway Act (1865; plan **99**) in Great Houghton, through Hardingstone, Northampton (St Giles and All Saints), terminating in a junction with the same railway in All Saints near where that railway is shown on the deposited plan as intended to cross Bridge Street (enlarged plan of town centre, also showing proposed new and altered roads near the railway).

Railway No 2. From a junction with Railway No 1 to a junction with the Midland Railway siding near the junction of that siding with the Blisworth and Peterborough branch of the London & North Western Railway, all in Hardingstone (Northants.).

Railway No 3. From a junction with Railway No 1 to a junction with the Blisworth and Peterborough branch of the LNWR near milepost 69, all in Hardingstone and Northampton (St Giles).

Railway No 4. From a junction with the authorised extension to Olney of the Newport Pagnell Railway (Newport Pagnell Railway (Extension to Olney) Act, 1865) to a junction with Railway No 1 as shown on the deposited plan referred to in the Bedford & Northampton Railway Act, 1865; all in Olney (Bucks.).

107. East & West Junction Railway. Hitchin Extension. James B. Burke, engineer. Deposited 30 Nov. 1865. No Act.

Railway No 1. From a junction in Towcester (Northants.) with the authorised line of the East & West Junction Railway (as per deposited plan [**94**] referred to in the East & West Junction Railway Act, 1864); through Paulerspury, Whittlebury, Potterspury, Furtho, Cosgrove (Northants.), Stony Stratford, Wolverton (Bucks.); terminating at Wolverton.

Railway No 2. From an end-on junction with Railway No 1 at Wolverton; through Haversham, Stantonbury, Great Linford, Little Linford, Newport Pagnell, Moulsoe (Bucks.); Cranfield, Salford, Holcot, Ridgmont, Husborne Crawley, Lidlington, and Millbrook (Beds.); terminating at Steppingley.

Railway No 3. From an end-on junction with Railway No 2 at Steppingley; through Flitwick, Flitton, Pulloxhill, Silsoe, Higham Gobion, Shillington (Beds.); Pirton and Ickleford (Herts.); terminating at Hitchin in a junction with the Great Northern Railway near the Hitchin Station of that company.

Railway No 4. From a junction with Railway No 2 at Steppingley; terminating in Flitwick (Beds.) in a junction with the authorised Midland Railway (as per plan referred to in Midland Railway (New Lines and Additional Powers) Act, 1864.

Railway No 5. From a junction with Railway No 1 at its terminus; terminating in a junction with the London & North Western Railway near Wolverton Station; all in Wolverton (Bucks.).

108. Newport Pagnell Railway. Extensions to Bedford & Northampton, Northampton & Peterborough and Leicester & Hitchin Railways. J.H. Tolmé, engineer. Deposited 30 Nov. 1867. Act 29 & 30 Vict. c. ccclix (1866).

Railway No 1. From a junction with the authorised extension to Olney of the Newport Pagnell Railway (Newport Pagnell Railway (Extension to Olney) Act, 1865) to a field near the junction between the Olney to Northampton and the Olney to Wellingborough roads, all in Olney (Bucks.).

Railway No 2. From an end-on junction with Railway No 1, through Warrington, Lavendon (Bucks.); Easton Maudit, Bozeat, Grendon, Strixton, Wollaston, Great Doddington,

Wellingborough, Irchester (Northants.); terminating at Irchester in a junction with the Northampton and Peterborough line of the London & North Western Railway.

Railway No 3. From the same starting-point as Railway No 1, terminating in a junction with the authorised line of the Bedford and Northampton Railway (Bedford & Northampton Railway Act, 1865; plan **99**), all in Olney.

Railway No 4. From the termination of Railway No 1, terminating in a junction with the Bedford and Northampton Railway, all in Olney.

Railway No 5. From a junction with Railway No 2 in Irchester, through Doddington, terminating at Wellingborough (Northants.) in a junction with the Leicester and Hitchin line of the Midland Railway near the Wellingborough Station on that line.

109. Peterborough Waterworks. George Bovill, engineer. Deposited 30 Nov. 1865. No Act.

Works in Wittering, Barnack, Thornhaugh, Wansford, Sutton, Castor, Marholm, Peterborough (Northants.).

110. Bedford & Northampton Railway. Charles Liddell, engineer. Deposited 30 Nov. 1866. Act 30 & 31 Vict. c. cxxiii (1867).

Railway No 1. From a junction in Clifton Reynes (Bucks.) with Railway No 1 authorised by the Bedford & Northampton Railway Act, 1865 [plan **99**]; through Lavendon, Warrington, Olney, Weston Underwood (Bucks.); Yardley Hastings (Northants.); Ravenstone (Bucks.); terminating in Weston Underwood (Bucks.) in a junction with Railway No 1.

Railway No 2. From a junction in Hardingstone (Northants.) with Railway No 2 authorised by the Bedford & Northampton Railway Act, 1866 [plan **108**]; through Northampton (St Giles and All Saints); terminating in a junction with Railway No 2 in All Saints (enlarged plan of Victoria Promenade and Bridge Street area; also plan of Midland Railway Goods Station).

111. Peterborough Waterworks. Ordish & Le Feuvre, T.S. Farrar, engineers. Deposited 30 Nov. 1866. Act 30 & 31 Vict. c. cxiv (1867).

Proposed waterworks in Deeping St James (Lincs.), with conduit to Peterborough through Peakirk, Maxey, Glinton, Werrington, Walton and Paston (Northants.).

112. Weedon & Daventry Railway. A.D. Johnstone, engineer. Deposited 30 Nov. 1867. Act 31 & 32 Vict. c. clxxi (1868).

From Daventry near the public pound on the road from Daventry to Weedon, through Newnham, Dodford, Norton, Brockhall, Weedon Beck; terminating in a junction with the London & North Western Railway near Weedon Station.

113. Peterborough Gas. Stears Brothers & Co., engineers. Deposited 29 Nov. 1867. Act 31 & 32 Vict. c. lxvi (1868).

Proposed gasworks on Westwood Road alongside Great Northern Railway in Peterborough.

114. Wellingborough Waterworks. George Smith, John Eunson, engineers. Deposited 30 Nov. 1867. No Act.

Plan shows a main from springs in a disused clay-pit alongside the turnpike road from Wellingborough to a reservoir at Great Harrowden, continuing to Hatton Garden, all in Wellingborough (Northants.).

115. Nene Valley Drainage and Navigation Improvements. John Fowler, engineer. Deposited 30 Nov. 1861. Act 25 & 26 Vict. c. clxiv (1862).

Works in Peterborough (Northants.); Fletton (Hunts.); Stanground (Cambs. & Hunts.); Whittlesey St Mary and St Andrew, Wisbech (Cambs.).

116. Midland Railway. Additional Powers. John Sydney Crossley, engineer. Deposited 29 Nov. 1867. Act 31 & 32 Vict. c. xliii (1868).

In Northants.: Additional land at Wellingborough, to the north of the station on the Leicester and Hitchin line, south of the Thrapston to Northampton road overbridge.

117. Thrapston Market. F.A. Hayward, surveyor. Deposited 30 Nov. 1869. Act 33 & 34 Vict. c. cxxxviii (1870).

Plan of land in Thrapston to be acquired for markets, corn exchange, slaughterhouse etc.

118. London & North Western Railway. Additional Powers. William Baker, engineer-in-chief; William Clarke, engineer. Deposited 30 Nov. 1868. Act 32 & 33 Vict. c. cxv (1869).

In Northants.: (a) New road and diversion of footpath in Thorpe Lubenham (Northants.) and Lubenham (Leics.); (b) New bridle road and occupation road in Lubenham (Leics.) and East Farndon (Northants.); both on the Rugby and Stamford line.

119. Newport Pagnell Railway. Alteration of Levels. J.H. Tolmé, engineer. Deposited 30 Nov. 1869. Act 33 & 34 Vict. c. xxix (1870).

Plan of portions of railway in Sherington, Tyringham with Filgrave, Emberton (Bucks.); and in Lavendon, Warrington (Olney) (Bucks.); Easton Maudit, Bozeat, Grendon, Strixton (Northants.); revising line authorised by Newport Pagnell Railway (Extension to Olney) Act, 1865, and Newport Pagnell Railway (Extension) Act, 1866.

120. Midland Counties & South Wales Railway. Extension to Buckinghamshire Railway. Charles Liddell, Edward Richards, engineers. Deposited 30 Nov. 1869. Act (Northampton & Banbury Junction Railway Act) 33 & 34 Vict. c. cxxii (1870).

From the existing railway in Bradden (as per deposited plan [**84**] of Northampton & Banbury Junction Railway, Act 1863), through Abthorpe, Slapton, Wappenham, Weedon Lois, Astwell with Falcutt, Helmdon, Stuchbury, Greatworth, Farthinghoe (Northants.); terminating in Greatworth in a junction with the Banbury Extension of the Buckinghamshire Railway. Described in *Gazette* notice as nearly identical with the incomplete portion of the railway which the company had power to construct under the Act of 1863.

120a. Northampton Cattle Market. E.F. Law, architect, surveyor to Northampton Corporation. Deposited 30 Nov. 1869. Act 33 & 34 Vict. c. xlv (1870).

Plan of proposed new markets in All Saints, St Giles, St Peter, St Sepulchre and St Andrew; and a proposed siding to the new cattle market from the authorised line of the Bedford & Northampton Railway, in All Saints.

121. Northampton Gas Light Co. Application under the Gas and Waterworks Facilities Act 1870 for provisional order authorising the raising of additional capital and for other purposes. Print of *Gazette* notice deposited 24 Nov. 1870. Twelve copies of the order filed in the office of the Clerk of the Peace 14 April 1871. No plan or book of reference. Order confirmed by Act 34 & 35 Vict. c. cxliv (1871).

122. Northampton Extension and Improvement. Plans and sections of proposed street improvements and sewage outfall and irrigation works. Lawson & Mansergh, consulting engineers. John Hyde Pidcock, engineer and surveyor to the Northampton Improvement Commissioners. Deposited 29 Nov. 1870. Act 34 & 35 Vict. c. cxxxix (1871).

Plan of land in Abington, Weston Favell, Little Billing, Great Billing, Ecton, Earls Barton, Northampton (All Saints) (Northants.), showing course of proposed outfall and site of sewage works.

123. Midland Railway. Additional Powers. John Sydney Crossley, engineer. Deposited 30 Nov. 1870. Act 33 & 34 Vict. c. lxiii (1870).

In Northants.: Additional land at Kettering Station and Rushton Station, both on the Leicester to London line (simple plans of both stations).

124. Weedon & Northampton Junction Railway. A.D. Johnstone, engineer. Deposited 30 Nov. 1870. No Act.

Railway No 1. From a junction with the authorised line of the Weedon & Daventry Railway (Act 1868; plan **112**) in Dodford, through Weedon Beck, Flore, Stowe Nine Churches, Upper Heyford, Kislingbury, Nether Heyford, Duston, Upton, Bugbrooke and Northampton (All Saints, St Peter, St Giles, St Andrew); terminating at the western end of the authorised Bedford & Northampton Railway (enlarged plan of built-up area in Northampton).

Railway No 2. From a junction with Railway No 1 to a junction with the London & North Western Railway near Weedon Station, all in Dodford.

125. Northampton & Banbury Junction Railway. Railway from Banbury to Blockley. Deviation in Authorised Extension to Ross. Charles Liddell, Edward Richards, engineers. Charles R. Cheffins, surveyor. Deposited 30 Nov. 1870. No Act.

Railway No 1. From a junction with the Great Western Railway at Blockley Staton in Blockley (Gloucs.); through Tidmington (Worcs.); Lower Lemington, Todenham, Stretton on Fosse (Gloucs.); Burmington, Great Wolford, Little Wolford, Long Compton, Stourton, Cherington, Whichford, Brailes, Sutton under Brailes (Warwicks.); Sibford Gower, Sibford Ferris, Hook Norton, Swalcliffe, Tadmarton, Broughton, Bloxham, Adderbury, Bodicote, Banbury, Neithrop (Oxon.); Warkworth, Middleton Cheney, Grimsbury (Northants.); terminating in Middleton Cheney in a junction with the Banbury Branch of the Buckinghamshire Railway of the LNWR near the Banbury Station on that branch.

Railway No 2. From a junction with Railway No 1 to a junction with the Banbury Branch near Banbury Station, all in Warkworth (Northants.) (good plan of both Banbury stations).

Deviation No 1. New line in lieu of portion of Railway No 3 authorised by the Northampton & Banbury Junction Railway Act, 1866, from a junction with the Ross & Monmouth Railway in Ross (Herefords.); through Weston under Penyard, Linton, Upton Bishop (Herefords.); Staunton (Gloucs.); Eldersfield (Worcs.); Oxenhall, Newent, Pauntley, Upleadon, Hartpury, Corse (Gloucs.); terminating at Corse in a junction with Railway No 3.

Deviation No 2. New line in lieu of portion of Railway No 3 as above from a junction with that railway in Chaceley (Gloucs.); through Bushley (Worcs.); Forthampton (Gloucs.); Tewkesbury (Worcs.); terminating in Tewkesbury by a junction with the Tewkesbury & Malvern Railway.

Deviation No 3. New line in lieu of portion of Railway No 1 authorised by 1866 Act, through Chipping Campden (Gloucs.).

126. Great Western Railway. Substituted Railway, New Railways, Roads, Additional Lands &c. W.G. Owen, Robert E. Johnston, engineers. Deposited 30 Nov. 1870. Act 34 & 35 Vict. c. clxxxiii (1871).

Includes a plan of a small area of additional lands alongside the Oxford to Banbury line in Kings Sutton (Northants.).

127. Coal Owners' Associated London Railway. Charles Bartholomew, civil engineer. Deposited 30 Nov. 1870. No Act.

Railway No 1. From Stratford le Bow (Middlesex), near the Lee Union Canal; through Hackney (Middlesex), Walthamstow (Essex); terminating in Leyton (Essex) in a junction with the Great Eastern Railway at Low Leyton Marsh.

Railway No 2. From a junction with the Great Eastern Railway in Trumpington (Cambs.); through Grantchester, Cambridge, Barton, Coton, Chesterton, Girton, Impington, Histon, Oakington, Westwick, Long Stanton St Michael, Long Stanton All Saints, Rampton, Willingham, Over (Cambs.); terminating in Willingham near Long Stanton Station on the St Ives and Cambridge branch of the Great Eastern Railway, there forming a junction with Railway No 3.

Railway No 3. From a junction with the St Ives and Cambridge branch of the Great Eastern Railway in Long Stanton All Saints (Cambs.); through Rampton, Over, Willingham (Cambs.); Bluntisham cum Earith, Colne, Somersham, Pidley cum Fenton, Warboys, Wistow, Bury, Ramsey (Hunts.); Stanground (Cambs. & Hunts.); Farcet (Hunts.); Peterborough, Paston, Gunthorpe, Walton, Werrington, Peakirk, Glinton, Northborough, Maxey, Deeping Gate (Northants.); Deeping St James, Market Deeping, Langtoft, Baston, Thurlby, Bourne (Lincs.); terminating in Bourne.

Railway No 4. From a junction with Railway No 3 in Bourne; through Edenham, Morton, Hacconby, Stainfield, Dunsby, Kirkby, Underwood, Rippingale, Dowsby, Aslackby, Millthorpe, Laughton, Folkingham, Sempringham, Birthorpe, Pointon, Billingborough, Horbling, Walcot, Threekingham, Spanby, Osbournby, Scredington, Burton Pedwardine, Silk Willoughby, Quarrington, Kirkby la Thorpe, Old Sleaford, New Sleaford (Lincs.); terminating in Kirkby Laythorpe in a junction with Railway No 5 near the Sleaford Station on the Sleaford to Bourn branch of the Great Northern Railway.

Railway No 5. From a junction with Railway No 4 in Kirkby la Thorpe; through Old Sleaford, Quarrington, New Sleaford, Holdingham, Leasingham, Roxholm, Cranwell, Ruskington, Brauncewell, Temple Bruer with Temple High Grange, Dorrington, Bloxholm, Digby, Ashby de la Launde, Rowlston, Scopwick, Blankney, Kirkby Green, Metheringham, Dunston, Nocton, Potter Hanworth, Branston, Washingborough, Heighington, Greetwell, Cherry Willingham, Branston (Lincs.); terminaing in Greetwell in a junction with the Market Raisen branch of the Manchester, Sheffield & Lincolnshire Railway.

128. Towcester & Hitchin Railway. James B. Burke, Joseph F. Burke, engineers. Deposited 30 Nov. 1871. No Act.

Railway No 1. From a junction with the authorised line of the East & West Junction Railway (Act, 1864; plan **94**) at Towcester; through Paulerspury, Whittlebury, Potterspury, Furtho, Cosgrove (Northants.); Stony Stratford, Wolverton, Haversham, Stantonbury, Great Linford, Little Linford,

Newport Pagnell, Moulsoe (Bucks.); Cranfield, Salford, Holcot, Ridgmont, Husborne Crawley, Lidlington, Millbrook, Steppingley, Flitwick, Flitton, Pulloxhill, Silsoe, Higham Gobion, Shillington (Beds.); Pirton, Ickleford (Herts.); terminating at Hitchin (Herts.) in a junction with the Royston and Hitchin branch of the Great Northern Railway.

Railway No 2. From a junction with Railway No 1 in Steppingley; terminating in Flitwick (Beds.) in a junction with the Midland Railway.

Railway No 3. From a junction with Railway No 1 to a junction with a siding of the London & North Western Railway near Wolverton Station, all in Wolverton (Bucks.).

129. Northampton & Banbury Junction Railway. Extension to Northampton. Charles Liddell, Edward Richards, engineers. Charles R. Cheffins, surveyor. Deposited 30 Nov. 1871. No Act.

From a junction with the branch railway authorised by the Northampton & Banbury Railway (Branch) Act, 1865 (plan 95), in Gayton; through Blisworth (good plans of Blisworth Station and Gayton Wharf on Grand Junction Canal), Milton, Wootton, Rothersthorpe, Kislingbury, Upton, Duston, Dallington, Northampton (St James's End, Cotton End, Far Cotton, St Peter, St Giles, All Saints, St Sepulchre, St Catherine, Priory of St Andrew, St Edmund) (Northants.); terminating in All Saints in a junction with the Bedford & Northampton Railway near the Northampton Station of that company (enlarged plan of area between Nene and station).

130. Daventry & Weedon Railway. A.D. Johnstone, engineer. Deposited 30 Nov. 1871. Act 35 & 36 Vict. c. clxxix (1872).

Railway No 1. From Daventry, through Newnham and Norton, terminating at Dodford (Northants.).

Railway No 2. From an end-on junction with Railway No 1, terminating in a junction with the London & North Western Railway, all within Dodford.

131. Kettering Waterworks. Thomas Hennell, engineer. Deposited 29 Nov. 1871. Provisional Order confirmed by Act 35 & 36 Vict. c. lxx (1872).

Plan showing pumping station in Weekley, reservoir in Kettering, and main pipes running through streets of town.

132. Midland Railway. Nottingham and Rushton Lines. John Sydney Crossley, engineer. Deposited 30 Nov. 1871. Act 35 & 36 Vict. c. cxviii (1872).

Railway No 1. Nottingham and Saxby Line. From a junction with the company's Nottingham and Lincoln line in Nottingham, through Sneinton, West Bridgford, Edwalton, Tollerton, Plumtree, Normanton on the Wolds, Stanton on the Wolds, Widmerpool, Willoughby on the Wolds, Hickling, Upper Broughton (Notts.); Nether Broughton, Old Dalby on the Wolds, Grimston, Saxelby, Asfordby, Welby, Sysonby, Melton Mowbray, Thorpe Arnold, Brentingby & Wyfordby, Freeby, Saxby (Leics.); terminating at Saxby in a junction with the company's Syston and Peterborough line near Saxby Station.

Railway No 2. Manton and Rushton Line. From a junction with the company's Syston and Peterborough line in Manton, through Wing, Preston, Ayston, Glaston, Bisbrooke, Uppingham, Liddington, Stoke Dry, Caldecott (Rutland); Stoke Dry, Great Easton, Drayton, Bringhurst, Holt (Leics.); Cottingham, East Carlton, Rockingham, Wilbarston, Great Oakley, Little Oakley, Rushton, Barford (Northants.); terminating in a junction with the company's main line to London near Rushton Station.

Railway No 3. Croxton Branch. From a junction with the intended Nottingham and Saxby line (as above) in Melton Mowbray, through Thorpe Arnold, Brentingby & Wyfordby, Waltham on the Wolds, Goadby Marwood, Stonesby, Croxton Keyrial, Eaton (Leics.); terminating at Waltham on the Wolds.

Railway No 4. Melton Branch. From a junction with the intended Nottingham and Saxby line (as above) in the township of Sysonby (Melton Mowbray), through Ab Kettleby, Welby, Melton Mowbray (Leics.); terminating at Melton Mowbray in a junction with the company's Syston and Peterborough line near Melton Mowbray Station on that line.

Railway No 5. Manton Curve. From a junction with the company's Syston and Peterborough line near Manton Station, terminating in a junction with the intended Manton and Rushton line (as above), all in Manton (Rutland).

133. Northampton & Daventry Junction Railway. A.D. Johnstone, engineer. Deposited 30 Nov. 1871. No Act.

From a junction with the proposed Daventry & Weedon Railway in Dodford, through Flore, Weedon Beck, Stowe Nine Churches, Upper Heyford, Kislingbury, Nether Heyford, Duston, Upton, Bugbrooke, Harpole, Northampton (All Saints, St Peter, St Giles, St Andrew) (Northants.); terminating in a junction with the Bedford and Northampton Railway in All Saints at their Northampton Station now in course of erection (enlarged plan of built-up area from gasworks to site of station).

134. East & West Junction Railway. James B. Burke, engineer. Deposited 30 Nov. 1872. No Act.

Railway No 1. From a junction with the Northampton & Banbury Railway near Blisworth Station in Blisworth parish, terminating nearby in Gayton (plans of Blisworth Station and Gayton Wharf).

Railway No 2. From a junction with Railway No 1; through Gayton, Blisworth, Rothersthorpe, Milton, Wootton, Kislingbury, Upton, Dallington, Duston, St James End, Cotton End, Far Cotton, Northampton (All Saints, St Peter, St Giles, St Sepulchre, St Katherine, St Andrew, St Edmund) (Northants.); terminating in a junction with the Bedford and Northampton Railway near that company's Northampton Station in All Saints (enlarged plan of built-up area from gasworks to station).

Railway No 3. From Byfield, through Charwelton and Hellidon; terminating in Hellidon (Northants.).

135. Midland and Manchester, Sheffield & Lincolnshire Railways. Extensions. W.H. Barlow, G.B. Bruce, engineers. Deposited 30 Nov. 1872. No Act.

Railway No 1. From a junction with the Midland main line to London in Barford (Northants.); through Rushton, Stoke Albany, Wilbarston, Ashley (Northants.); Medbourne, Blaston, Hallaton, East Norton, Loddington, Tilton, Halstead, Marefield, Owston & Newbold, Burrough on the Hill, Twyford, Great Dalby, Melton Mowbray, Sysonby, Welby, Holwell, Ab Kettleby, Nether Broughton, Long Clawson (Leics.); Colston Basset (Notts.); Hose (Leics.); Langar cum Barnston, Cropwell Bishop, Cropwell Butler, Bingham, Shelford, East Bridgford, Gunthorpe, Lowdham, Caythorpe, Gonalston, Epperstone, Oxton, Farnsfield, Bilsthorpe, Rufford, Edwinstowe, Ollerton, Perlethorpe cum Budby, Bothamsall, Worksop (Notts.); terminating in a junction with the main line of the MS&LR.

Railway No 2. From a junction with Railway No 1 in Babworth; through Barnby Moor, Hodsock, Blyth, Styrrup (Notts.); terminating in Tickhill (Yorks.).

Railway No 3. From an end-on junction with Railway No 2;

through Tickhill, Wadworth, Loversall, Balby with Hexthorpe, Doncaster, Carr House & Elmfield, Cantley, Wheatley, Bentley with Arksey and Thorpe in Balne (Yorks.); terminating in Owston in a junction with the North Eastern Railway near Shaftholme Junction.

Railway No 4. From an end-on junction with Railway No 2; through Tickhill, Wadworth, Doncaster (Yorks.); terminating in a junction with the railway of the South Yorkshire Railway and the River Dun Company.

Railway No 5. From a junction with the Rugby and Stamford branch of the London & North Western Railway in Weston by Welland (Northants.); through Ashley (Northants.); Medbourne (Leics.); terminating in a junction with Railway No 1.

Railway No 6. From a junction with Railway No 1 in Melton Mowbray (Leics.); through Sysonby; terminating in a field in Melton Mowbray.

Railway No 7. From a junction with Railway No 1 in Lowdham; through Caythorpe (Notts.); terminating in Lowdham in a junction with the Nottingham and Lincoln line of the Midland Railway.

Railway No 8. From a junction with the Nottingham and Lincoln line of the Midland Railway in Lowdham; through Caythorpe; terminating in Lowdham in a junction with Railway No 1.

Railway No 9. From a junction with Railway No 1, terminating in a junction with the Mansfield and Southwell branch of the Midland Railway, all in Farnsfield (Notts.).

Railway No 10. From a junction with the MS&LR main line in Worksop (Notts.); terminating in a junction with Railway No 2, also in Worksop.

Railway No 11. From a junction with the South Yorkshire Railway Doncaster to Thorne line, terminating in a junction with Railway No 3, all in Bentkey with Arksey (Yorks.)

Railway No 12. From a junction with Railway No 3, terminating in a junction with the South Yorkshire Railway Doncaster to Thorne line, all in Bentley with Arksey (Yorks.)

Railway No 13. From a junction with Railway No 3 in Owston; through Thorpe in Balne, Burghwallis (Yorks.); terminating in Owston in a junction with the Lancashire & Yorkshire Railway.

Railway No 14. From a junction with Railway No 4 in Doncaster, terminating in a junction with the railway of the South Yorkshire Railway and River Dun Company, all in Doncaster.

Railway No 15. From a junction with the main line of the MS&LR in Worksop (Notts.); through North & South Anston, Dinnington, Laughton en le Morthen, Maltby, Stainton with Hellaby, Braithwell; Conisbrough (Yorks.); terminating in a junction with the railway of the South Yorkshire Railway and River Dun Company.

Railway No 16. From a junction with Railway No 15 in Anston (Yorks.); through Todwick, Wales, Treeton (Yorks.); terminating in Todwick in a junction with the MS&LR.

136. Midland Railway. Additional Powers. John Sydney Crossley, engineer. Deposited 29 Nov. 1872. Act 36 & 37 Vict. c. ccx (1873).

In Northants.: Widening of existing main line between Rushton and Bedford, through Barford, Rothwell, Glendon, Kettering, Pytchley, Barton Seagrave, Burton Latimer, Isham, Little Harrowden, Great Harrowden, Finedon, Wellingborough, Irthlingborough, Irchester (Northants.); Wymington, Souldrop, Sharnbrook, Felmersham, Milton Ernest, Pavenham, Oakley, Bromham, Biddenham, Bedford (Beds.).

137. Market Harborough, Melton Mowbray & Nottingham Railways. William Baker, John Fraser, engineers. Deposited 29 Nov. 1872. No Act.

Railway No 1. From a junction with the termination of Railway No 3 authorised by the Great Northern Railway (Newark and Melton) Act, 1872, at Melton Mowbray, through Welby, Sysonby, Brentingby & Wyfordby, Asfordby, Thorpe Arnold, Kirby Bellars, Frisby on the Wreak, Burton Lazars, Great Dalby, Gaddesby, Ashby Pasture, Thorpe Satchville, Pickwell with Leesthorne, Burrough on the Hill, Somerby, Twyford, South Croxton, Marefield, Owston & Newbold, Lowesby, Cold Newton, Halstead, Tilton, Billesdon, Skeffington, Withcote, Whatborough, Loddington, Tugby, East Norton, Allexton, Goadby, Noseley, Horninghold, Stockerston, Hallaton, Glooston, Cranoe, Stonton Wyville, Slawston, Blaston, Holt, Medbourne, Great Easton, Bringhurst, Drayton, Thorpe Langton, Welham (Leics.); Cottingham, East Carlton, Ashley, Weston by Welland, Sutton Basset (Northants.); terminating in a junction with the Rugby and Stamford branch of the London & North Western Railway in Weston.

Railway No 2. From a junction with Railway No 1 in Hallaton, through Slawston and Medbourne (Leics.), terminating in a junction with the Rugby and Stamford branch of the LNWR in Ashley (Northants.).

Railway No 3. From a junction with the Nottingham and Grantham branch of the Great Northern Railway in Saxondale; through Shelford, Saxondale, Bingham, Cropwell Butler, Cropwell Bishop, Wiverton Hall, Whatton, Elton, Langar cum Barnston, Granby (Notts.); Plungar, Harby, Stathern (Leics.); terminating in a junction with Railway No 2 authorised by the 1872 Act (as above) in Stathern.

138. Banbury & Cheltenham Direct Railway. E. Wilson, engineer. Deposited 30 Nov. 1872. Act 36 & 37 Vict. c. clxxii (1873).

Railway No 1. From a junction with the Birmingham and Oxford line of the Great Western Railway near Kings Sutton Station in Kings Sutton (Northants.); through Adderbury, Milton, Bloxham, Milcombe, Wigginton, Hook Norton, Great Rollright, Over Norton; terminating in a junction with the Chipping Norton line of the GWR near Chipping Norton Station in Chipping Norton (Oxon.).

Railway No 2. From a junction with Railway No 1 near its commencement, through Kings Sutton and Adderbury, to a junction with the Birmingham and Oxford line of the GWR near Kings Sutton Station.

Railway No 3. From a junction with the Chipping Norton line of the GWR in Churchill, near Chipping Norton Junction Station; through Kingham (Oxon.), terminating in Bledington (Gloucs.) in a junction with the Bourton on the Water Railway near Chipping Norton Junction Station.

Railway No 4. From a junction with the Bourton on the Water Railway in Bourton on the Water; through Lower Slaughter, Upper Slaughter, Cold Aston, Naunton, Guiting Power, Notgrove, Salperton, Compton Abdale, Sevenhampton, Shipton, Whittington, Dowdeswell, Charlton Kings, Leckhampton; terminating at Cheltenham (Gloucs.) in a junction with the railway between Cheltenham and Gloucester.

Railway No 5. From a junction with Railway No 4 in Cheltenham; terminating in a junction with the railway between Cheltenham and Gloucester, also in Cheltenham.

139. Great Northern Railway. Additional Powers. Richard Johnson, engineer. Deposited 29 Nov. 1872. Act 36 & 37 Vict. c. xc (1873).

In Northants.: Additional lands alongside main line at (a) Peterborough, in Fletton (Hunts.) and Peterborough, on either side of the river Nene; and (b) between Peterborough and New England.

140. Blockley & Banbury Railway. Charles Liddell, Edward Richards, engineers. Charles R. Cheffins, surveyor. Deposited 30 Nov. 1872. No Act.

Railway No 1. From a junction with the Great Western Railway at Blockley Station in Blockley (Gloucs.); through Tidmington (Warwicks.); Lower Lemington, Todenham, Stretton on Fosse (Gloucs.); Burmington, Great Wolford, Little Wolford, Long Compton, Stourton, Cherington, Whichford, Brailes, Sutton under Brailes (Warwicks.); Sibford Gower, Sibford Ferris, Hook Norton, Swalcliffe, Tadmarton, Milcombe, Broughton, Bloxham, Adderbury, Bodicote, Banbury, Neithrop (Oxon.); Warkworth, Middleton Cheney, Grimsbury (Northants.); terminating in Warkworth in a junction with the Banbury branch of the Buckinghamshire Railway of the London & North Western Railway near Banbury Station on that branch.

Railway No 2. From a junction with Railway No 1 to a junction with the Banbury Branch, near Banbury Station, all in Warkworth (Northants.).

Railway No 3. From a junction with Railway No 2 to a junction with the GWR near that company's Banbury Station, all in Warkworth.

Railway No 4. From a junction with Railway No 1 to a junction with Railway No 2, all in Warkworth.

141. London & North Western Railway. New Lines. William Baker, engineer-in-chief. Francis Stevenson, engineer. Deposited 28 Nov. 1872. Act 36 & 37 Vict. c. clvi (1873).

In Northants.:

Seaton and Wansford Railway. From a junction with the company's Rugby and Stamford line in Seaton (Rutland) near Seaton Station, through Barrowden (Rutland); Wakerley, Fineshade, Blatherwycke, Kings Cliffe, Rockingham, Apethorpe, Nassington (Northants.); Elton, Sibson cum Stibbington (Hunts.); terminating there in a junction with the company's Northampton and Peterborugh branch.

New Railway at Kelmarsh. Deviation at Kelmarsh Tunnel on the Northampton and Market Harborough line, including land in Kelmarsh and Arthingworth (Northants.).

New Railway at Oxendon. Deviation at Oxendon Tunnel on the Northampton and Market Harborough line, entirely in Great Oxendon.

142. Great Northern and London & North Western Railways. New Lines from Market Harborough and Stathern to Bingham. William Baker, John Fraser, engineers. Deposited 28 Nov. 1873. Act 37 & 38 Vict. c. clvii (1874).

Railway No 1. From a junction with Railway No 2 authorised by the Great Northern Railway (Melton to Leicester) Act, 1873, at its termination in Tilton (Leics.), through Billesdon, Skeffington, Withcote, Whatborough, Loddington, Tugby, East Norton, Allexton, Goadby, Noseley, Horninghold, Stockerston, Hallaton, Glooston, Cranoe, Stonton Wyville, Slawston, Blaston, Holt, Medbourne, Great Easton, Bringhurst, Drayton, Thorpe Langton, Welham (Leics.); Cottingham, East Carlton, Ashley, Weston by Welland, Sutton Basset (Northants.); terminating in a junction with the Rugby and Stamford branch of the LNWR in Weston.

Railway No 2. From a junction with Railway No 1 in Hallaton, through Blaston, Holt, Slawston, Bringhurst, Great Easton, Drayton, Medbourne (Leics.); Ashley, East Carlton, Cottingham (Northants.); terminating in Bringhurst (Leics.) in the Rugby and Stamford branch of the LNWR.

Railway No 3. From a junction with the Nottingham and Grantham branch of the Great Northern Railway in Saxondale (Notts.), through Shelford, Bingham, Cropwell Butler, Cropwell Bishop, Wiverton Hall, Whatton, Elton, Langar, Barnston cum Granby (Notts.); Plungar, Harby, Stathern (Leics.); terminating in a junction with Railway No 2 authorised by the 1872 Act (as above) in Stathern.

143. Peterborough Gas Works. Deposited 29 Nov. 1873. Act 37 & 38 Vict. c. xxxvi (1874).

Plan of lands in Peterborough, adjoining St John's Street and Thorney Road.

144. London & North Western Railway. England and Ireland. William Baker, engineer-in-chief; Francis Stevenson, engineer. Deposited 29 Nov. 1873. Act 37 & 38 Vict. c. clix (1874).

In Northants.:

New Railway at Kelmarsh. Deviation on Northampton and Market Harborough line at Kelmarsh Tunnel, including land in Kelmarsh and Arthingworth (Northants.).

New Railway at Oxendon. Deviation on Northampton and Market Harborough line at tunnel in Great Oxendon (Northants.).

145. Midland Railway. Additional Powers. Kettering and Manton Line. William Henry Barlow, John Sidney Crossley, engineers. Deposited 29 Nov. 1873. Act 37 & 38 Vict. c. cxl (1874).

In Northants.:

Kettering and Manton Line. From a junction with the company's main line to London in Barford, through Kettering, Rushton, Glendon, Newton, Geddington, Great Oakley, Little Oakley, Stanion, Corby, Great Weldon, Little Weldon, Rockingham, Deene, Deenethorpe, Gretton, Harringworth, Laxton (Northants.); Seaton, Thorpe by Water, Morcott, Glaston, Liddington, Bisbrooke, Uppingham, Wing, Preston, Pilton, Lyndon, Martinsthorpe, Manton (Rutland); terminating in Manton in a junction with the company's Syston and Peterborough line near Manton Station.

146. Great Northern Railway. Further Powers. John Fraser, Richard Johnson, engineers. John Fowler, consulting engineer. Deposited 29 Nov. 1873. Act 37 & 38 Vict. c. clviii (1874).

In Northants.: Additional lands in (a) in Peterborough; (b) Helpston.

147. London & North Western Railway. New Lines and Additional Powers. William Baker, engineer-in-chief. Francis Stevenson, engineer. Deposited 28 Nov. 1874. Act 38 & 39 Vict. c. clii (1875).

In Northants.: (a) New road and road to be stopped up in Wakerley on the authorised Seaton and Wansford line; (b) Additional lands to be purchased in Hardingstone, at Bridge Street Station on the Northampton to Peterborough branch (simple plan of station).

148. East & West Junction Railway. Blisworth Branch. Joseph F. Burke, engineer. Deposited 30 Nov. 1874. No Act.

From a junction with the Northampton and Banbury Railway in Gayton near Blisworth Station; terminating in Blisworth (Northants.) in a junction with the Northampton and Peterborough branch of the London & North Western Railway (plans of Blisworth Station and Gayton Wharf).

149. London & North Western Railway. Bletchley, Northampton and Rugby Railway. William Baker, engineer-in-chief. Francis Stevenson, engineer. Deposited 27 Nov. 1874. Act 38 & 39 Vict. c. cii (1875).

Railway No 1. Widening of existing main line through Bletchley, Woughton on the Green, Loughton, Bradwell, Bradwell Abbey, Wolverton, Haversham, Castlethorpe, Hanslope (Bucks.); Hartwell, Ashton, Roade, Courteenhall, Milton, Wootton, Hardingstone, Duston, Northampton (St Peter, All Saints), Dallington (Northants.); terminating in a junction with the company's Northampton and Market Harborough branch near Castle Station.

Railway No 2. From a junction with the company's Northampton and Market Harborough branch in Kingsthorpe, through Dallington, Harlestone, Church Brampton, Holdenby, Althorp, Brington, East Haddon, Long Buckby, Watford, Crick, Kilsby, Barby (Northants.); Hillmorton, Clifton on Dunsmore, Rugby (Warwicks.); terminating in a junction with the company's main line.

Railway No 3. From a junction with Railway No 1 to a junction with the company's Northampton and Peterborough branch, all in Hardingstone (Northants.).

150. Peterborough Water Co. Shelford & Robinson, engineers. Deposited 30 Nov. 1874. No Act.

Plan includes land in Peterborough, Paston, Werrington, Glinton, Etton, Northborough, Maxey (Northants.), Market Deeping, Langtoft, Baston, Greatford, Wilsthorpe, Braceborough (Lincs.), showing course of conduit from storage tank in Braceborough to the town.

151. Buckinghamshire & Northamptonshire Railways Union Railway. Edward Wilson, engineer. Deposited 30 Nov. 1874. No Act.

Railway No 1. From a junction with the Aylesbury and Buckingham Railway in East Claydon (Bucks.); through Middle Claydon, Steeple Claydon and Padbury (Bucks.); terminating in Buckingham.

Railway No 2. From an end-on junction with Railway No 1 at Buckingham; through Buckingham, Radclive, Stowe, Westbury (Bucks.); Brackley (Northants.); Luffield Abbey (Bucks.); Whitfield, Silverstone, Abthorpe (Northants.), terminating at Abthorpe.

Railway No 3. From an end-on junction with Railway No 2 at Abthorpe; terminating in Bradden (Northants.) in a junction with the Northampton & Banbury Junction Railway.

Railway No 4. From a junction with Railway No 2 at Abthorpe; through Bradden and Greens Norton; terminating in Towcester (Northants.) in a junction with the East & West Junction Railway.

Railway No 5. From a junction with Railway No 4 at Abthorpe; through Bradden, Greens Norton, Wood Burcote; terminating in Greens Norton (Northants.) in a junction with the East & West Junction Railway.

152. Northampton & Blisworth Railway. George Hopkins, engineer. Deposited 30 Nov. 1874. No Act.

Railway No 1. From a junction with the Bedford & Northampton Railway at that company's Northampton terminus, through the parishes of All Saints, St Peter (enlarged plan of built-up area between station and gasworks), Dallington, Duston (Northants.).

Railway No 2. From a junction with Railway No 1, through All Saints, St Peter, Dallington, Hardingstone, Duston, Upton, Kislingbury, Wootton, Rothersthorpe, Milton, Blisworth and Gayton; terminating at Gayton (Northants.).

Railway No 3. From a junction with Railway No 2, terminating in a junction with the Northampton & Banbury Junction Railway, entirely in Gayton.

Railway No 4. From a junction with Railway No 3, terminating in Blisworth in a junction with the Northampton and Peterborough branch of the London & North Western Railway.

Railway No 5. From a junction with Railway No 1, through Northampton (St Peter, All Saints), Dallington, Hardingstone and Duston, terminating in a junction with the Northampton and Market Harborough branch.

153. London & North Western Railway. Additional Powers. William Baker, engineer-in-chief. Francis Stevenson, engineer. Deposited 29 Nov. 1875. Act 39 & 40 Vict. c. clxxx (1876).

In Northants.:

Additional Land and Buildings at Pitsford. Land in Pitsford, alongside Northampton to Market Harborough branch, with plan of Brampton Station.

Additional Land at Thorpe Achurch. Land in Thorpe Achurch, alongside Northampton to Peterborough branch, with plan of Thorpe Achurch Station.

New Road at Weston. On the authorised line of the Tilton to Market Harborough line (Great Northern & LNWR Joint Lines Act, 1874), close to the LNWR Rugby to Stamford line.

154. Midland Railway. New Works &c. Rushton and Bedford Widening Deviation. Messrs Barlow, Son & Baker, engineers. Deposited 30 Nov. 1875. Act 39 & 40 Vict. c. cxlv (1876).

In Northants.:

Rushton and Bedford Widening Deviation. From a junction in Irthlingborough on the company's railway from Rushton to Bedford as authorised to be widened by the Midland Railway (Additional Powers) Act, 1873; through Irchester (Northants.); Wymington, Souldrop, Sharnbrook and Felmersham (Beds.); terminating in Milton Ernest in a junction with the same railway.

155. Duston Minerals & Northampton & Gayton Junction Railway. Charles Bartholomew, engineer. Deposited 30 Nov. 1875. No Act.

From a junction with the Northampton & Banbury Junction Railway in Gayton, through Blisworth, Rothersthorpe, Milton, Wootton, Kislingbury, Upton, Duston, Northampton (St Peter, All Saints) (Northants.); terminating in All Saints in a junction with the Bedford & Northampton Railway near that company's Northampton terminus (enlarged plan of built-up area from gasworks to station).

156. Braceborough Water. W. Shelford, engineer. Deposited 30 Nov. 1875. No Act.

Lands in Braceborough, Wilsthorpe, Greatford, Baston, Langtoft, Market Deeping (Lincs.); Maxey, Northborough, Etton, Glinton, Werrington, Paston (Northants.).

157. Midland Railway. New Works. County of Northampton. John Underwood, engineer. Deposited 30 Nov. 1876. Act 40 & 41 Vict. c. lii (1877).

Market Harborough Loop. From a junction with the company's main line in Great Bowden (Leics.) to a junction in

Little Bowden (Northants.).

Cransley Branch. From a junction with the company's main line near Kettering Station (plan), through Kettering, Boughton and Cransley; terminating at Cransley (Northants.).

New Road at Gretton. In Gretton on the company's Kettering and Manton line (in course of construction).

Aqueduct at Peterborough. In Peterborough, on the river Nene south of the company's Syston to Peterborough line.

Additional Land at Ufford. In Ufford, on the company's Syston to Peterborough line.

158. Great Northern Railway. John Fraser, Richard Johnson, engineers. Deposited 29 Nov. 1876. Act 40 & 41 Vict. c. lxxx (1877).

Additional lands at New England, Peterborough (Northants.), to the north of the engine shed there.

159. London & North Western Railway. New Works and Additional Lands. William Baker, engineer-in-chief. Francis Stevenson, engineer. Deposited 29 Nov. 1876. Act 40 & 41 Vict. c. xlv (1877).

In Northants.: New footpath crossing the Seaton and Wansford line (in course of construction) in Kings Cliffe, near a bridge carrying the road from Kings Cliffe to Duddington over the railway.

160. Great Eastern Railway. Northern Extension. Sir John Hawkshaw, engineer. Deposited 30 Nov. 1877. Act 41 & 42 Vict. c. xc.

Railway No 1. From a junction with the company's Cambridge and St Ives railway near Long Stanton Station in Long Stanton All Saints (Cambs.); through Willingham, Over (Cambs.); Bluntisham cum Earith, Colne, Somersham, Pidley cum Fenton, Warboys, Wistow, Bury, Ramsey (Hunts.); Whittlesey St Mary & St Andrew, Thorney (Cambs.); Eye, Newborough, Borough Fen (Northants.); Crowland, Deeping St Nicholas, Spalding, Cowbit (Lincs.); terminating in Spalding on the bank of the river Welland.

Railway No 2. From an end-on junction with Railway No 1; through Spalding, Pinchbeck, Gosberton, Surfleet, Quadring, Donington, Helpringham, Burton Pedwardine, Kirkby la Thorpe, New Sleaford, Leasingham, Ruskington, Dorrington, Digby, Rowston, Kirkby Green, Timberland, Scopwick, Blankney, Metheringham, Dunston, Nocton, Potter Hanworth, Branston, Washingborough, Canwick, Boultham, Skellingthorpe, Lincoln, Burton by Lincoln (Lincs.); terminating in Burton by Lincoln.

Railway No 3. From an end-on junction with Railway No 2; through Burton by Lincoln, South Carlton, North Carlton, Scampton, Aisthorpe, Brattleby, Cammeringham, Ingham, Fillingham, Glentworth, Harpswell, Hemswell, Willoughton, Great Corringham, Pilham, Blyton, Laughton, Owston Ferry, Haxey, Epworth, Wroot (Lincs.); Hatfield, Fishlake, Barnby Dun, Owston, Burghwallis (Yorks.); terminating in a junction with the Lancashire & Yorkshire Railway in Burghwallis.

Railway No 4. From a commencement in Colne, through Somersham (Hunts.), terminating in Colne in a junction with the company's St Ives and March line.

Railway No 5. From a junction with Railway No 1 to a junction with the company's Peterborough and Ely railway, all in Whittlesey St Mary & St Andrew (Cambs.).

Railway No 6. From a junction with Railway No 1, terminating in a junction with the Peterborough, Wisbeach & Sutton Bridge Railway, all in Thorney (Cambs.).

Railway No 7. From a junction with Railway No 1 in Spalding, through Cowbit (Lincs.), terminating in Spalding in a junction with the March and Spalding line of the Great Northern Railway.

Railway No 8. From a junction with Railway No 2 in Pinchbeck (Lincs.), terminating in Spalding in a junction with the loop line of the GNR from Peterborough to Spalding and Boston.

Railway No 9. From a junction with Railway No 2 at its termination; through Burton by Lincoln, South Carlton, Broxholme, Saxilby cum Ingleby, Torksey; terminating in Torksey in a junction with the Manchester, Sheffield & Lincolnshire Railway.

Railway No 10. From a junction with Railway No 3 in Pilham, terminating in a junction with the MS&LR in Blyton (Lincs.).

Railway No 11. From a junction with Railway No 3 in Barnby Dun (Yorks.), terminating in a junction with the South Yorkshire Line of the MS&LR in the same parish.

Railway No 12. From a junction with Railway No 3 in the township of Thorpe in Balne; through Barnby Dun, Burghwallis, Campsall, Owston (Yorks.), terminating in a junction with the York and Doncaster Branch of the North Eastern Railway.

161. Daventry & Weedon Railway. A.D. Johnstone, engineer. Deposited 30 Nov. 1877. No Act.

From a field near the road from Daventry to Norton, in Daventry, through Newnham, Norton, Dodford (Northants.); terminating in Dodford in a junction with the London & North Western Railway near Weedon Station.

162. Market Deeping Railway. Walrond Smith, engineer. Deposited 29 Nov. 1877. Act 41 & 42 Vict. c. lxxxvii (1878); abandonment Act 46 & 47 Vict. c. clxx (1883).

From a junction with the main up line of the Great Northern railway in Helpston; through Etton, Glinton, Maxey, Deeping Gate, Northborough (Northants.); terminating in Market Deeping (Lincs.) adjoining the main road from Market Deeping to Stamford.

To be constructed under the Regulation of Railways Act, 1868, on the light railway system.

163. London & North Western Railway. Additional Powers. William Baker, engineer-in-chief. Francis Stevenson, engineer. Deposited 30 Nov. 1877. Act 41 & 42 Vict. c. clxxxi (1878).

In Northants.: (a) New bridle road crossing the Northampton to Rugby line (in course of construction) in Long Buckby; (b) New bridle road crossing the Seaton to Wansford line (in course of construction) in Kings Cliffe; (c) Diversion of river Nene alongside the Northampton to Market Harborough line to the north of West Bridge and Castle Station, in Northampton St Peter, Dallington and Northampton St Sepulchre.

164. Midland Railway. Additional Powers. John Underwood, engineer. Deposited 30 Nov. 1877. Act 41 & 42 Vict. c. xcvi (1878).

In Northants.: (a) Two new footpaths in Kettering, both crossing the Leicester to London line, one near the Rothwell to Kettering road and the other near the Wellingborough to Kettering road. The sheet of plans also includes a plan showing a new footpath at Manton (Rutland), which includes a plan of Manton Station on the company's Peterborough to Syston line. (b) Additional land in St Mark, Peterborough, to the west of the company's line between Spital Bridge and Westwood Bridge.

165. Midland Railway. Additional Powers. John Underwood, engineer (thus on cover; Barlow, Son & Baker sign the plan). Deposited 29 Nov. 1878. Act 42 & 43 Vict. c. cviii (1879).

In Northants.: New footpath at Glendon and Barford, alongside the Leicester to London line.

166. Wellingborough Gas Works. Application by Wellingborough Gas Light Co. Ltd for a Provisional Order under the Gas and Waterworks Facilities Act, 1870. Dated 11 Nov. 1878. No memorial. Order confirmed by Act 42 & 43 Vict. c. clix (1879).

Plan shows layout of existing gasworks alongside main road from Newport Pagnell to Wellingborough, identifying individual rooms and plant.

167. London & North Western Railway. Additional Powers. William Baker, engineer-in-chief. Francis Stevenson, engineer. Deposited 29 Nov. 1878. Act 42 & 43 Vict. c. cxlii (1879).

In Northants.: (a) Diversion of bridle road at Roade Station (good plan of station) on the London to Rugby line; (b) Diversion of bridle road and new footpath on the Northampton to Rugby line (in course of construction) in Harlestone; (c) New footpath on the same line in Brington; (d) New footpath on the same line in Long Buckby; (e) New road, proposed extension of bridge, additional lands and buildings, in Little Bowden near the junction of the Northampton to Market Harborough and Rugby to Market Harborough lines; (f) New footpaths in Catthorpe (Leics.) and Lilbourne (Northants.) near Lilbourne Station on the Rugby to Stamford line; (g) additional lands and buildings in St Peter, Northampton, near Castle Station, to the south of West Bridge.

168. Daventry & Weedon Railway. A.D. Johnstone, engineer. Deposited 30 Nov. 1878. No Act.

From a field near the road from Daventry to Norton, in Daventry, through Newnham, Norton, Dodford (Northants.); terminating in Dodford in a junction with the London & North Western Railway near Weedon Station. (Identical to **161**).

169. Easton Neston Mineral & Towcester, Roade & Olney Junction Railway. J.H. Shipway, L.H. Shirley, engineers. Deposited 30 Nov. 1878. Act 42 & 43 Vict. c. ccxxiii (1879).

Railway No 1. From a junction with the East & West Junction Railway west of Towcester Station in Towcester, to a junction with Railway Nos 3 and 4 in Easton Neston, forming a flying junction between the E&WJR and Railway No 4.

Railway No 2. From a junction with the Northampton & Banbury Junction Railway east of Towcester Station to a junction with Railway No 3, all in Towcester.

Railway No 3. From a junction with the Northampton & Banbury Junction Railway east of the commencement of Railway No 2 to a junction with Railway Nos 1 and 4 in Easton Neston. (Railway Nos 2 and 3 form a triangular junction with the N&BJR.)

Railway No 4. From an end-on junction with Railway Nos 1 and 3 in Easton Neston, through Shutlanger, terminating in Stoke Bruerne.

Railway No 5. From an end-on junction with Railway No 4, through Stoke Bruerne, Roade, Quinton, Courteenhall, Preston Deanery, Piddington, Horton, terminating in Horton in a south-facing junction with the Bedford and Northampton Railway.

Railway No 6. From a junction with Railway No 5 to a terminus alongside the London & North Western Railway at Roade Station, all in Roade.

Railway No 7. From a junction with Railway No 5 to form a north-facing junction with the Bedford and Northampton Railway, in Horton.

Railway No 8. From a junction with Railway No 5 to a terminus alongside the Grand Junction Canal near the first overbridge south of the southern entrance to Blisworth Tunnel, all in Stoke Bruerne.

170. London & North Western Railway. Francis Stevenson, engineer. Deposited 28 Nov. 1879. Act 43 & 44 Vict. c. cxlv (1880).

In Northants.: (a) Additional lands in Duddington, near Top Lodge Farm, alongside the Seaton and Wansford line; (b) Additional lands in Kings Cliffe on the same line.

171. Northampton Tramways. Thomas Floyd, engineer. Deposited 29 Nov. 1879. Act 43 & 44 Vict. c. ci (1880).

Tramway No 1. Commencing in Hardingstone near the entrance to the London & North Western Railway Station at Northampton, along London Road and Bridge Street, terminating in All Saints at the junction of Bridge Street and Gold Street.

Tramway No 2. From the end of Tramway No 1, along the Drapery, the Parade, Sheep Street, Regent Square, Royal Terrace and Barrack Road, terminating in St Sepulchre near the junction of Barrack Road and St George's Terrace.

Tramway No 3. Commencing in High Street (in Duston or Dallington) near the junction of that road with the road to Weedon, along High Street, West Bridge, Black Lion Hill, Mare Fair and Gold Street, terminating in a junction with Tramway No 1.

Tramway No 4. Commencing in All Saints at the junction of Mercers Row and the Drapery, along Mercers Row, Abington Street, Abington Square and Kettering Road, terminating in St Giles in Kettering Road near the junction of that road and the road leading to Kingsthorpe.

Tramway No 5. Commencing in a junction with Tramway No 4 at the junction of Kettering Road and Wellingborough Road, along Wellingborough Road, terminating in St Giles at the junction of Wellingborough Road and East Street.

Tramway No 6. Commencing in All Saints in a junction with Tramway No 4 near the north-east corner of All Saints churchyard, along Wood Hill, St Giles Square, St Giles Street, Spencer Parade and Billing Road, terminating in St Giles near the junction of Billing Road and Upper Thrift Street.

Tramway No 7. Commencing in a junction with Tramway No 2 in All Saints, at the junction of the Drapery and the Parade, along the Parade, the north and east sides of the Market Square, terminating in the same parish in a junction with Tramway No 4 near the south-east corner of the Market Square.

Works in Hardingstone, Duston, Dallington, St James, All Saints, St Sepulchre, St Peter, St Giles, St Andrew, St Mary, and St Katherine. Gauge 3 ft 6 in. Numerous passing places shown on plans. All lines to be single except that part of Tramway No 1 between the junction of Bridge Street and Cattle Market Road and the junction of Bridge Street with the road leading to the Midland Railway Station, which will be double.

172. Peterborough Tramways. John Addy, engineer. Application for a Provisional Order under the Tramways Act 1870. Deposited 29 Nov. 1879. Order confirmed by Act 43 & 44 Vict. c. clxxii (1880).

Tramway No 1. Commencing in Lincoln Road near the junction with Crown Street, along Lincoln Road, Boroughbury, Westgate, Long Causeway, Market Place, Church Street, Cowgate, terminating in Cowgate near the Great Northern Railway Station approach road. Single track with passing places.

Tramway No 2. From a junction with Tramway No 1 in Lincoln Road, near the junction with Cobden Street, along Lincoln Road, Lincoln Road East, Monument Street, Cemetery End, New Road, Mid Gate and Long Causeway, terminating in a junction with Tramway No 1 in Long Causeway near Mid Gate. Single track with passing places.

Tramway No 3. From a junction with Tramway No 1 in Lincoln Road near Cobden Street, terminating in a junction with Tramway No 2 in Lincoln Road East near Henri Street. Single track.

Tramway No 4. From a junction with Tramway No 1 in Westgate, opposite the west side of Long Causeway, and terminating by a junction with Tramway No 2 in Mid Gate, opposite the east side of Long Causeway. Single track.

All in Peterborough. Gauge 3 ft 6 in.

173. London & North Western Railway and Midland Railway. Market Harborough New Line and Works. Francis Stevenson, John Underwood, engineers. Deposited 29 Nov. 1880. Act 44 & 45 Vict. c. xlvii (1881).

Deviation of Midland main line in Great Bowden (Leics.) near the junction with the LNWR Rugby and Stamford branch, with a diversion of the main road in Great Bowden village, at which point the line of deviation rejoins the existing LNWR/MR line. The line of deviation is then shown continuing alongside the existing line through Market Harborough Station, terminating at the second overbridge on the Midland line north of the station, and including a detailed plan of the station.

174. Midland Railway. Additional Powers. John Underwood, engineer. Deposited 29 Nov. 1880. Act 44 & 45 Vict. c. cli (1881).

In Northants.: Additional lands in Isham, alongside the Leicester to London main line, between Finedon Furnaces and Finedon Station (neither of which is planned).

175. London & North Western Railway. New Railways. Francis Stevenson, engineer. Deposited 27 Nov. 1880. Act 44 & 45 Vict. c. cxli (1881).

In Northants.:

Addington Branch. From a junction with the company's Northampton and Peterborough line in Woodford near Thrapston, through Great Addington, Finedon and Burton Latimer, terminating in Burton Latimer.

Islip Branch. From a junction with the company's Northampton and Peterborough line, terminating in a junction with the sidings of Islip Ironworks (planned), all in Islip.

Northampton and Kingsthorpe Widening. On the Northampton and Rugby line in Dallington and Kingsthorpe.

176. Daventry & Weedon Railway. R.J.H. Saunders, engineer. Deposited 30 Nov. 1880. No Act.

From a field near the road from Daventry to Norton, in Daventry, through Newnham, Norton, Dodford (Northants.); terminating in Dodford in a junction with the London & North Western Railway near Weedon Station. See **161, 168**.

177. Kettering Gas. Application by Kettering Gas Co. Ltd for a provisional order under the Gas and Waterworks Facilities Act, 1870. Deposited 30 Nov. 1881. Order confirmed by Act 45 & 46 Vict. c. xcix (1882).

Plan of the existing works, showing location and use of individual plant and buildings.

178. Midland Railway. Additional Powers. John Underwood, engineer. Deposited 30 Nov. 1881. Act 45 & 46 Vict. c. cxxx (1882).

In Northants.: Small piece of additional land at Uffington Station in Barnack, adjoining the Great Northern Railway Stamford to Wansford line.

179. Northampton Tramways. Extensions. Thomas Floyd, engineer. Deposited 30 Nov. 1881. Act 45 & 46 Vict. c. cx (1882).

Tramway No 1. From a junction with Tramway No 2 authorised by Northampton Street Tramways Act 1880, in St Sepulchre near the junction of Marriott Road and Barrack Road, along Barrack Road and Kingsthorpe Road, terminating in Kingsthorpe near the junction of Kingsthorpe Road and the road to Brampton. Single track with passing places.

Tramway No 2. From a junction with Tramway No 3 authorised by 1880 Act in Duston near the junction of Devonshire Street and St James End, along St James End, terminating in the road leading to Duston near the Melbourne Gardens and the junction of Argyll Road with the road leading to Duston. Single track with one passing place.

Works in Northampton St Sepulchre, Kingsthorpe and Duston. See **171**.

180. Northampton Corporation. Commons, Streets, Roads Bill. John Hyde Pidcock, Borough Surveyor. Deposited 30 Nov. 1881. Act 45 & 46 Vict. c. ccxii (1882).

Commons. Plans of commons to be acquired by the Corporation, viz. New Commons (St Giles), Peaches Meadow (St Giles), Midsummer Meadow (St Giles), Cow Meadow (St Giles and All Saints), Baulms Holme (All Saints), Foot Meadow (St Peter, All Saints), Gates Meadow (St Andrew), Race Course (St Sepulchre, St Giles).

New Road. Plan of proposed new road between St John's Street and Victoria Promenade in All Saints.

New Street and Widening. Plan of proposed new street between Pike Lane and Horse Market, and widening of Marefair, in All Saints parish (very detailed).

181. London & North Western Railway. Francis Stevenson, engineer. Deposited 30 Nov. 1881. Act 45 & 46 Vict. c. cxxix (1882).

In Northants.: (a) Additional lands in Kingsthorpe, between the river Nene and Northampton to Rugby line, near junction with Market Harborough branch; (b) Additional lands in Harlestone, alongside the Northampton to Rugby line and the road from Church Brampton to Harlestone; (c) Additional lands in Holdenby, alongside the Northampton to Market Harborough branch at Althorp Station (good plan of station).

182. Midland Railway. Additional Powers. John Underwood, engineer. Deposited 29 Nov. 1882. Act 46 & 47 Vict. c. cxi (1883).

In Northants.: New footpath in Irchester at Irchester Station on the Leicester to Bedford line (good plan of station).

183. Great Northern Railway. Richard Johnson, John Wilson, engineers. John Fowler, consulting engineer. Deposited 30 Nov. 1883. Act 47 & 48 Vict. c. xxvi (1884).

In Northants.: Subway to be constructed to carry footway (only) beneath the level crossing by which the Great Northern and Midland lines cross Thorpe Road in Peterborough; carriageway unaffected. Detailed plan of adjoining portion of Great Northern station.

184. London & North Western Railway. Francis Stevenson, engineer. Deposited 29 Nov. 1883. Act 47 & 48 Vict. c. ccvii (1884).

In Northants.: (a) Additional lands in Nether Heyford, alongside London to Birmingham line at the bridge over the road from Nether Heyford to Stowe, and adjoining Heyford Ironworks (exchange sidings planned, but not ironworks); (b) Additional lands at Buckby Bank in Long Buckby, in the angle between the London to Birmingham line and the Grand Junction Canal, adjoining the road from Daventry to Long Buckby; (c) Extension of bridge on the Rugby and Stamford line over the road from Great Bowden to Dingley, and additional lands, in Little Bowden (Northants.) and Great Bowden (Leics.), near Market Harborough Station.

185. Northampton & Daventry Railway. R. Elliott Cooper, Leane & Bakewell, engineers. Deposited 30 Nov. 1883. No Act.

Railway No 1. From a junction with the terminus of the Northampton & Bedford Railway in All Saints (Northampton); through Northampton (St Peter, St James, St Giles, All Saints, St Andrew, Abbey Walls) (enlarged plan of town between station and gasworks, over which the railway was to pass on a viaduct 1,056 yards long), Dallington, Duston, Upton, Kislingbury, Harpole, Bugbrooke, Nether Heyford, Upper Heyford, Flore, Stowe Nine Churches, Weedon Beck, Brockhall, Dodford, Norton, Newnham, Daventry; terminating in Daventry at the road from Daventry to Weedon.

Railway No 2. From a junction with Railway No 1 in Dodford, terminating in the same parish in a junction with the London & North Western Railway.

186. Northampton Water. T. & C. Hawksley, engineers. Deposited 30 Nov. 1883. Act 47 & 48 Vict. c. ccviii (1884).

Plan of proposed Ravensthorpe Reservoir, in the parishes of Ravensthorpe and Guilsborough; pipe from reservoir through Ravensthorpe, Holdenby, Spratton, Chapel Brampton, Boughton and Kingsthorpe, terminating in St Giles (Northampton) at an existing covered service reservoir of the Northampton Water Works Co. known as the Kettering Road Reservoir.

187. Stratford-upon-Avon, Towcester & Midland Junction Railway. Deviation. Charles Liddell, engineer. Deposited 29 Nov. 1884. Act 48 & 49 Vict. c. cxliii (1885).

From a junction with the Northampton & Banbury Junction Railway in Towcester (Northants.), terminating in Easton Neston (Northants.) in a junction with authorised Railway No 4 of the Easton Neston Mineral & Towcester, Roade & Olney Junction Railway (Act, 1879; plan **169**).

188. London & North Western Railway. Francis Stevenson, engineer. Deposited 28 Nov. 1884. Act 48 & 49 Vict. c. lxxxviii (1885).

In Northants.:

Weedon & Daventry Railway. From a junction with the company's London and Birmingham railway in Dodford, through Newnham, terminating in Daventry at the road from Daventry to Long Buckby.

Weedon Deviation. In the parishes of Weedon Beck and Dodford, deviation of the London and Birmingham railway between the south end of Weedon Viaduct and a point to the north of Weedon Station, easing curve through station (planned in detail).

189. Northampton & Banbury & Metropolitan Junction Railway. No engineer named. Deposited 28 Nov. 1884. No Act.

Railway No 1. From a junction with the Aylesbury & Buckingham Railway in East Claydon (Bucks.); through Middle Claydon, Steeple Claydon and Padbury (Bucks.); terminating in Radclive (Bucks.) on the south bank of the river Ouse.

Railway No 2. From an end-on junction with Railway No 1; through Stowe, Biddlesden (Bucks.); Syresham (Northants.); Luffield Abbey (Bucks.); Whitfield, Silverstone and Abthorpe (Northants.); terminating in Towcester (Northants.) in a junction with the Northampton & Banbury Junction Railway near Greens Norton Junction.

Railway No 3. From a junction with Railway No 1 at its termination, to a junction with the Banbury and Buckingham branch of the London & North Western Railway, all in Radclive.

Railway No 4. From a junction with the Northampton & Banbury Junction Railway in Gayton (Northants.), terminating in Blisworth in a junction with the Northampton and Peterborough branch of the LNWR.

190. Northampton, Daventry & Leamington Railway. Wells-Owen & Elwes, and Leane & Bakewell, engineers. Deposited 29 Nov. 1884. No Act.

Railway No 1. From a junction with the Northampton & Bedford Railway at its Northampton terminus in All Saints; through Northampton (St Peter) (enlarged plan of built-up area as in **185**, including proposed viaduct of 1,056 yards), Duston, Upton, Kislingbury, Bugbrooke, Upper Heyford, Nether Heyford, Flore, Stowe Nine Churches; terminating in Dodford (Northants.).

Railway No 2. From an end-on junction with Railway No 1 in Dodford, terminating in the same parish in a junction with London & North Western Railway.

Railway No 3. From an end-on junction with Railway No 1 in Dodford; through Newnham, Norton; terminating in Daventry.

Railway No 4. From an end-on junction with Railway No 3 in Daventry; through Braunston (Northants.); Wolfhampcote, Grandborough, Napton on the Hill, Stockton, Southam, Long Itchington, Ufton, Offchurch, Radford Semele (Warwicks.); terminating in Leamington Priors (Warwicks.).

Railway No 5. From an end-on junction with Railway No 4, terminating in a junction with the Rugby to Leamington line of the LNWR, all in Leamington Priors.

Railway No 6. From an end-on junction with Railway No 4, terminating in a junction with the Oxford to Leamington line of the Great Western Railway, all in Leamington Priors.

191. Kettering Water Works. Thomas Hennell, engineer. Application for a provisional order under the Gas and Waterworks Facilities Act 1870. Deposited 28 Nov. 1885. Order confirmed by Act 49 & 50 Vict. c. lix (1886).

Plan of lands in Grafton Underwood, Warkton and Weekley (Northants.), showing pumping engine, existing engine house and well, and conduits.

192. Wolverton & Stony Stratford Tramways. Deanshanger Extension. Stephen Sellon, engineer. Application by the Wolverton & Stony Stratford Tramways Co. Ltd for a provisional order under the Tramways Act 1870. Deposited 30 Nov. 1886. Order confirmed by Act 51 & 52 Vict. c. cxxiii (1887).

From an end-on junction with Tramway No 11 authorised by the Wolverton & Stony Stratford Tramways Order, 1883, in High Street, Stony Stratford (Bucks.), along High Steet, crossing the Ouse into Old Stratford Road in Cosgrove and Passenham parishes (Northants.); along Deanshanger Road in Passenham to a terminus at Deanshanger village, with a branch diverging shortly before the terminus and running to the entrance to Messrs E. & H. Roberts's Works at Deanshanger.

Single track with passing places. Gauge 3 ft 6 in.

193. Midland Railway. Additional Powers. John Underwood, engineer. Deposited 30 Nov. 1887. Act 51 & 52 Vict. c. cxviii (1888).

In Northants.: New footpath and additional lands in Isham at Isham Station (plan) on the London and Leicester line.

194. Kettering Waterworks. Thomas Hennell, engineer. Application by Kettering Waterworks Co. Ltd for a provisional order under the Gas and Waterworks Facilities Act 1870. Deposited 29 Nov. 1887. Order confirmed by Act 51 & 52 Vict. c. xvii (1888).

Plan of intended new pumping station and tank, and additional lands at existing reservoir in Kettering.

195. Kettering Water. Thomas Hennell, engineer. Deposited 30 Nov. 1888. Act 52 & 53 Vict. c. xix (1889).

Plan of proposed reservoir in the parishes of Loddington, Cransley and Thorpe Malsor, with a pipe from the reservoir to an intended service reservoir on the western edge of Kettering parish, and from there into the town along Lower Street and Northall Street, terminating in a junction with the existing main pipe on Rockingham Road and Newland Street.

196. Worcester & Broom Railway. Extension to Aylesbury. Charles Liddell, Edward Richards, engineers. Deposited 30 Nov. 1888. No Act.

Railway No 1. From a junction with the East & West Junction Railway in Canons Ashby (Northants.); through Moreton Pinkney, Sulgrave, Helmdon, Wappenham, Radstone, Brackley St Peter (Northants,); Turweston (Bucks.); Evenley (Northants.); Westbury (Bucks.); Mixbury, Finmere (Oxon.); Tingewick, Barton Hartshorn, Chetwode, Preston Bisset, Twyford, Steeple Claydon, Grendon Underwood, Quainton (Bucks.); terminating in a junction with the Aylesbury & Buckingham Railway in Quainton, near Quainton Road Station.

Railway No 2. From a junction with Railway No 1 in Quainton; through Waddesdon, Fleet Marston, Quarrendon, Aylesbury and Hartwell (Bucks.); terminating in a junction with the Aylesbury & Buckingham Railway in Aylesbury.

Railway No 3. From a junction with the East & West Junction Railway, terminating in a junction with Railway No 1, all in Moreton Pinkney (Northants.).

Railway No 4. From a junction with the Banbury branch of the London & North Western Railway in Brackley St Peter; terminating in a junction with Railway No 1 in Turweston (Bucks.).

197. London & North Western Railway. Francis Stevenson, engineer. Deposited 29 Nov. 1888. Act 52 & 53 Vict. c. xcviii (1889).

In Northants.: Additional lands at Wellingborough Station (good plan) in Irchester.

198. Wellingborough & District Tramroads. Stephen Sellon, engineer. Deposited 30 Nov. 1888. Act 52 & 53 Vict. c. ccxii (1889).

Tramroad No 1. From London Road, Bozeat, near the junction with Mile Street, through Wollaston, terminating near Wellingborough Station on the LNWR opposite the Prince of Wales public house.

Tramroad No 2. From Sheep Street, Wellingborough, near the Bees Wing public house, past the gasworks, over the river Nene and LNWR, terminating in a junction with Tramroad No 1.

Tramroad No 3. From near the gasworks in Rushden on the north side of Church Street, near its junction with Alfred Street, along High Street, Higham Ferrers, terminating at Higham Ferrers Station on the LNWR.

Tramroad No 4. From an end-on junction with Tramroad No 3, along the road leading to Stanwell and along the Old Roman Road, thence towards Raunds, terminating in Brook Street there, opposite the Wesleyan Methodist Chapel.

Tramroad No 4a. Short passing loop at junction of Tramroads 3 and 4.

In Bozeat, Strixton, Wollaston, Irchester, Wellingborough, Higham Ferrers, Rushden, Chelveston cum Caldecott, Stanwick and Raunds (Northants.). Gauge 3 ft 6 in.

199. Towcester & Buckingham Railway. No engineer named. Deposited 28 Nov. 1888. Act 52 & 53 Vict. c. cci (1889).

Railway No 1. From a junction with the Aylesbury & Buckingham Railway in East Claydon (Bucks.); through Middle Claydon, Steeple Claydon, Padbury and Buckingham (Bucks.); terminating in Radclive (Bucks.) on the south bank of the river Ouse.

Railway No 2. From an end-on junction with Railway No 1; through Stowe, Biddlesden (Bucks.); Syresham (Northants.); Luffield Abbey (Bucks.); Whittlebury and Silverstone (Northants.); terminating in Towcester (Northants.).

Railway No 3. From a junction with Railway No 1, to a junction with the Banbury and Buckingham branch of the LNWR, all in Radclive.

Railway No 4. From a junction with Railway No 2 in Towcester, terminating in Easton Neston (Northants.) in a junction with the railway authorised by the Stratford-upon-Avon, Towcester & Midland Junction Railway Act 1885 (plan **187**).

Railway No 5. From a junction with Railway No 2, terminating in a junction with the Northampton & Banbury Junction Railway, all in Towcester.

200. Northampton Electric Lighting. 1:10,560 Ordnance Survey map showing proposed area of supply, endorsed 'Board of Trade 1889'. Lodged in the office of Clerk of the Peace 2 July 1889. Area includes Northampton and parts of Abington, Hardingstone, Wootton, Duston, Dallington and Kingsthorpe. Order confirmed by Act 53 & 54 Vict. c. cxcvi (1890). See **206**.

201. Midland Railway. Alfred A. Langley, engineer. Deposited 28 Nov. 1889. Act 53 & 54 Vict. c. cxxxviii (1890).

In Northants.:

Irchester and Raunds Branch. From a junction with the company's Leicester to Bedford line in Irchester; through Rushden, Higham Ferrers, Chelveston cum Caldecott, Stanwick;

terminaing in Raunds in a junction with the company's Kettering, Thrapston and Huntingdon branch near Raunds Station. Plan includes enlarged detail of brickworks at Raunds.

202. London & North Western Railway. Francis Stevenson, engineer. Deposited 30 Nov. 1889. Act 53 & 54 Vict. c. cliv (1890).

In Northants.:

Daventry and Leamington Railway. From a junction with the company's Weedon and Daventry railway at its terminus at Daventry Station (planned) in Daventry; through Braunston (Northants.); Wolfhampcote, Grandborough, Leamington Hastings, Stockton (two lime and cement works alongside canal planned in detail), Birdingbury, Long Itchington (lime and cement works planned), Hunningham (Warwicks.); terminating in a junction with the company's Rugby and Leamington railway.

Additional Land at Thorpe Station. Land in Thorpe Achurch on Northampton to Peterborough line (station planned).

203. Metropolitan Railway. Extension to Moreton Pinkney. Charles Liddell, engineer. Deposited 30 Nov. 1889. Act 53 & 54 Vict. c. cxxciii (1890).

Railway No 1. From a junction with the East & West Junction Railway in Canons Ashby; through Moreton Pinkney, Sulgrave, Helmdon, Wappenham, Radstone, Brackley (Northants.); Turweston, Westbury (Bucks.); Evenley (Northants.); Mixbury, Finmere, Shelswell, Newton Purcell (Oxon.); Barton Hartshorn, Chetwode (Bucks.); Godington (Oxon.); Twyford, Steeple Claydon, Grendon Underwood; terminating in Quainton (Bucks.) in a junction with the Aylesbury & Buckingham Railway.

Railway No 2. From a junction with Railway No 1 in Quainton; through Waddesdon, Fleet Marston, Quarrendon, Hartwell; terminating in Aylesbury (Bucks.) in a junction with the Aylesbury & Buckingham Railway.

Railway No 3. From a junction with the East & West Junction Railway, terminating in a junction with Railway No 1, all in Moreton Pinkney (Northants.).

Railway No 4. From a junction with the Northampton & Banbury Junction Railway near Helmdon Station, to a junction with Railway No 1, all in Helmdon.

Railway No 5. From a junction with the Northampton & Banbury Junction Railway near Helmdon Station, to a junction with Railway No 1, all in Helmdon.

Railway No 6. From a junction with the Banbury branch of the LNWR in Evenley (Northants.); terminating in Mixbury (Oxon.) in a junction with Railway No 1.

204. Great Northern Railway. Various Powers. Richard Johnson, John Fraser & Sons, engineers. Sir John Fowler, consulting engineer. Deposited 29 Nov. 1889. Act 53 & 54 Vict. c. civ (1890).

In Northants.: Additional lands near Werrington Junction, in Paston, adjoining the company's main line and the road from Bourn to Marholm.

205. Wellingborough & District Tramroads. Extensions. Stephen Sellon, engineer. Deposited 30 Nov. 1889. Act 53 & 54 Vict. c. ccxiv (1890).

Tramroad No 1. From a junction in London Road, Wellingborough, with Tramroad No 2, authorised by the Wellingborough and District Tramroads Act, 1889 (plan **198**), near the junction with Cemetery Road, along Cemetery Road and Midland Road, terminating opposite the Midland Hotel.

Tramroad No 2. From a junction in Midland Road with Tramroad No 2, along Midland Road to the Midland Railway Station, Wellingborough, terminating there.

Tramroad No 3. From a junction in London Road with Tramroad No 1 authorised by the 1889 Act, near the junction with Gipsy Lane, along Gipsy Lane, through Irchester, terminating in the road leading to the Midland Railway Station at Irchester.

Tramroad No 4. From an end-on junction with Tramroad No 3, along the bridge over the Midland Railway, terminating 4 chains from its commencement.

Tramroad No 5. From an end-on junction with Tramroad No 4, past the Oakley public house, along the road leading to Higham Ferrers, terminating in a field near the gasworks, Rushden, on the north side of Church Street near its junction with Alfred Street.

Tramroad No 6. From a junction with Tramroad No 5 in the road leading to Higham Ferrers near the stream situated between the junction of the roads leading to Higham Ferrers, Rushden and Irchester Station, and the Oakley Arms public house, along Rushden Hill and High Street, Higham Ferrers, terminating in the Market Place, Higham Ferrers, near the Town Hall.

Tramroad No 7. From a junction in London Road, Bozeat, with Tramroad No 1 authorised by the 1889 Act, near the Chequers public house, across the road from Bozeat to Easton Maudit, past Bozeat Mill, thence towards Olney into the London Road, terminating on the north side of the Midland Railway bridge near the junction of the Bedford Road and the London Road.

Tramroad No 8. From an end-on junction with Tramroad No 7, thence under the Midland Railway bridge, terminating in the Olney Road in a junction with Tramway No 12 authorised by the Newport Pagnell and District Tramways Order, 1887, near the Queens Hotel, Olney.

In the parishes of Wellingborough, Irchester, Rushden, Higham Ferrers and Bozeat (Northants.), and Lavendon and Olney (Bucks.).

To be built on gauge of 3 ft 6 in., with power to lay additional rail on gauge of 4 ft 8½ in.

206. Northampton Electric Light & Power Co. Ltd. Application for a provisional order under the Electric Lighting Acts 1882 and 1888, lodged with clerk of the peace 28 Nov. 1889. With a 1:10,560 Ordnance Survey map showing same area of supply as **200**. Order confirmed by Act 53 & 54 Vict. c. cxcvi (1890).

207. Midland Railway. New Lines &c. J.A. McDonald, engineer. Deposited 28 Nov. 1890. Act 54 & 55 Vict. c. xl (1891).

In Northants.: Additional land at Peterborough, west of the company's line, north of Mayor's Walk, near the engine shed (partly planned).

208. Great Northern Railway. Richard Johnson, John Fraser & Sons, engineers. Sir John Fowler Bt, consulting engineer. Deposited 28 Nov. 1890. Act 54 & 55 Vict. c. xix (1891).

In Northants.: Widening from Werrington Junction to Helpston on the company's main line through Werrington, Marholm, Glinton, Etton and Helpston.

209. Manchester, Sheffield & Lincolnshire Railway. Extension to London Etc. Charles Liddell, engineer-in-chief.

Edward Richards, Edward Parry, engineers. Deposited 27 Nov. 1890. No Act.

Railway No 1. From a junction in Kirkby in Ashfield (Notts.) with Railway No 9 authorised by the MSLR Act of 1889; through Annesley, terminating in a branch with the Great Northern Railway (Leen Valley Branch) in Newstead (Notts.).

Railway No 2. From a junction in Kirkby in Ashfield with Railway No 1; through Newstead, Linby, Hucknall Torkard, Bulwell, Basford, Radford, Nottingham (enlarged plan of town centre), South Wilford, Ruddington, Gotham, East Leake, Stanford upon Soar (Notts.); Prestwold, Loughborough, Barrow upon Soar, Swithland, Thurcaston, Newtown Linford, Beaumont Leys, Anstey Pastures, Gilroes, Leicester (enlarged plan of town centre) (Leics.) terminating in Bridge Street, Leicester.

Railway No 3. From the termination of Railway No 2; through Leicester, Glenfield, Aylestone, Whetstone, Cosby, Ashby Magna, Dunton Bassett, Gilmorton, Lutterworth, Misterton, Cotesbach, Shawell (Leics.); terminating in Clifton on Dunsmore (Warwicks.).

Railway No 4. From the termination of Railway No 3; through Clifton on Dunsmore, Rugby, Hillmorton (Warwicks.); Barby (Northants.); Willoughby (Warwicks.); Braunston (Northants.); Wolfhampcote, Upper Shuckburgh (Warwicks.); Catesby, Hellidon, Charwelton (Northants.); terminating in Woodford cum Membris (Northants.).

Railway No 5. From the termination of Railway No 4; through Woodford cum Membris, Eydon, Canons Ashby, Moreton Pinkney, Sulgrave, Helmdon, Wappenham, Radstone, Brackley (Northants.); Turweston, Westbury (Bucks.); Evenley (Northants.); Mixbury, Finmere, Shelswell, Newton Purcell (Oxon.); Barton Hartshorn, Chetwode (Bucks.); Godington (Oxon.); Twyford, Steeple Claydon, Grendon Underwood (Bucks.); terminating in Quainton in a junction with the Aylesbury & Buckingham Railway of the Metropolitan Railway near Quainton Road Station.

Railway No 6. From the termination of Railway No 3, terminating in a junction with the Rugby and Peterborough branch of the LNWR near Clifton Mill Station, all in Clifton on Dunsmore.

Railway No 7. From the termination of Railway No 4, terminating in a junction with the East & West Junction Railway, all in Woodford cum Membris.

Railway No 8. From a junction with Railway No 7 to a junction with Railway No 5, all in Woodford cum Membris.

Railway No 9. From a junction with the Metropolitan Railway in Hampstead near West Hampstead Station, terminating in St Marylebone (Middlesex) (on enlarged scale).

Railway No 10. A widening of the Metropolitan Railway in Willesden (Middlesex) and Hampstead (on enlarged scale).

Railway No 11. From a junction with Railway No 10 near Kilburn & Brondesbury Station, to a junction with the Hampstead Junction Branch of the LNWR, near West End Lane Station, all in Hampstead (on enlarged scale).

Railway No 12. From a junction with Railway No 9 authorised under the Act of 1889, terminating in a junction with the private railway of the Pinxton Colliery Co., all in Kirkby in Ashfield.

Railway No 13. From a junction with Railway No 1 in Kirkby in Ashfield, through Sutton in Ashfield, Skegby, Teversall (Notts.); Ault Hucknall (Derbys.); terminating in Pleasley (Derbys.) near Pleasley Colliery.

Railway No 14. From a junction with Railway No 13 in Teversall (Notts.), terminating elsewhere in the same parish.

Deviation No 1. From a junction with Railway No 9 authorised by the 1889 Act, terminating in a junction with the same railway, all in Sutton cum Duckmanton (Derbys.).

Deviation No 2. From a junction with Railway No 9 authorised by the 1889 Act, terminating in a junction with the same railway, all in Tibshelf (Derbys.).

Deviation No 3. From a junction with Railway No 9 in Blackwell (Derbys.), terminating in Sutton in Ashfield (Notts.).

The deviations are not included in book of plans, but are scheduled in the book of reference.

210. Crowland Railway. James Wilkinson, engineer. Deposited 29 Nov. 1890. Act 54 & 55 Vict. c. lxxxviii (1891).

From a junction with the Great Northern Railway (Lincolnshire Loop Line) in Peakirk (Northants.), through Newborough and Borough Fen (Northants.), terminating in Crowland (Lincs.).

To be laid to either standard gauge or a narrower gauge (of not less than 3 ft); if laid to narrow gauge, provision for a third rail to be laid along GNR standard gauge line from junction to Peakirk Station.

211. London & North Western Railway. Additional Powers. Francis Stevenson, engineer. Deposited 28 Nov. 1890. Act 54 & 55 Vict. c. cxxxvii (1891).

In Northants.: Diversion of road in Higham Ferrers and Chelveston cum Caldecott, at Higham Ferrers Station (planned) on the Northampton to Peterborough line.

212. Midland Railway. J.A. McDonald, engineer. Deposited 30 Nov. 1891. Act 55 & 56 Vict. c. ccxxxiv (1892).

In Northants.: (a) New footpath at Higham Ferrers, alongside Irchester and Raunds branch (under construction); (b) New footpath at Irchester, on the same line.

213. Manchester, Sheffield & Lincolnshire Railway. Extension to London Etc. Charles Liddell, engineer-in-chief. Edward Richards, Edward Parry, engineers. Deposited 27 Nov. 1891. Act 56 & 57 Vict. c. i (1893).

Railway No 1. From a junction in Kirkby in Ashfield with Railway No 1 authorised by the MSLR Act of 1891; through Newstead, Linby, Hucknall Torkard, Bulwell, Basford, Radford, Nottingham (enlarged plan of town centre), South Wilford, Ruddington, Gotham, East Leake, Normanton upon Soar (Notts.); Prestwold, Loughborough, Barrow upon Soar, Swithland, Thurcaston, Rothley, Wanlip, Birstall, Belgrave, Leicester (enlarged plan of town centre) (Leics.), terminating in Applegate Street.

Railway No 2. From the termination of Railway No 1; through Leicester, Glenfield, Aylestone, Lubbersthorpe, Whetstone, Cosby, Ashby Magna, Dunton Bassett, Gilmorton, Lutterworth, Misterton, Cotesbach, Shawell (Leics.); terminating in Clifton on Dunsmore (Warwicks.).

Railway No 3. From the termination of Railway No 2; through Clifton on Dunsmore, Rugby, Hillmorton (Warwicks.); Barby (Northants.); Willoughby (Warwicks.); Braunston (Northants.); Wolfhampcote, Upper Shuckburgh (Warwicks.); Catesby, Hellidon, Charwelton (Northants.); terminating in Woodford cum Membris (Northants.).

Railway No 4. From the termination of Railway No 3; through Woodford cum Membris, Eydon, Canons Ashby, Moreton Pinkney, Sulgrave, Helmdon, Wappenham, Radstone, Brackley (Northants.); Turweston, Westbury (Bucks.); Evenley (Northants.); Mixbury, Finmere, Shelswell, Newton Purcell (Oxon.); Barton Hartshorn, Chetwode (Bucks.); Godington (Oxon.); Preston Bisset, Twyford, Steeple Claydon, Grendon Underwood (Bucks.); terminating in Quainton in a junction with the Aylesbury & Buckingham Railway of the Metropolitan Railway near Quainton Road Station.

Railway No 5. From the termination of Railway No 2, terminating in a junction with the Rugby and Peterborough branch of the LNWR near Clifton Mill Station, all in Clifton on Dunsmore.

Railway No 6. From the termination of Railway No 3, terminating in a junction with the East & West Junction Railway, all in Woodford cum Membris.

Railway No 7. From a junction in Woodford cum Membris with the East & West Junction Railway to a junction with Railway No 4 in Eydon (Northants.).

Railway No 8. From a junction with the Metropolitan Railway in Hampstead, near West Hampstead Station, terminating in St Marylebone (Middlesex) at Nos 222-224 Marylebone Road.

Railway No 9. A widening of the Metropolitan Railway in the Willesden (Middlesex) and Hampstead.

Railway No 10. From a junction with Railway No 9 near Kilburn & Brondesbury Station, to a junction with the Hampstead Junction Branch of the LNWR, near West End Lane Station, all in Hampstead.

Railway No 11. From a junction with Railway No 8 to a point on the north side of Princess Street near the junction with Carlisle Street, all in St Marylebone.

Railway No 12. From a junction with Railway No 8 at Alpha Road, to a junction with the Metropolitan Railway near Baker Street Station, all in St Marylebone.

Further lines (Nos 13–20) included in the *Gazette* notice are not included in the Northamptonshire plans or book of reference. The plans for Railways Nos 8–11 above are on an enlarged scale.

214. Northampton Tramways. Julius Gottschalk, engineer. Application to Board of Trade for a provisional order. Deposited 28 Nov. 1891. With a second copy marked 'Clerk of the Council's Copy'. Order confirmed by Act 55 & 56 Vict. c. cxciv (1892).

Tramway from a junction with the existing tramway of the Northampton Street Tramways Co. in Abington Square, near the junction of Kettering Road and Wellingborough Road, in Northampton St Giles, along Wellingborough Road, terminating in that road in Abington near Abington Abbey.

Single track with passing places. Gauge 3 ft 6 in.

215. Midland Railway. J.A. McDonald, engineer. Deposited 29 Nov. 1892. Act 56 & 57 Vict. c. lviii (1893).

In Northants.: (a) Additional lands at Wellingborough, in Wellingborough, Great Harrowden and Finedon parishes, near Wellingborough Ironworks, alongside the Leicester to Bedford line; (b) Additional lands at Helpston, in Bainton, alongside the Syston to Peterborough line.

216. London & North Western Railway. Francis Stevenson, engineer. Deposited 29 Nov. 1892. Act 56 & 57 Vict. c. clxvi (1893).

In Northants.: (a) Junction at Weedon, wholly in Dodford, from the company's London and Birmingham railway and the Weedon and Daventry railway (plan of Weedon Station); (b) Deviation at Braunston on the Daventry and Leamington railway, in Braunston (Northants.) and Wolfhampcote (Warwicks.); (c) Additional lands and buildings at Roade, Courteenhall and Milton, alongside the Bletchley, Northampton and Rugby railway; (d) Additional lands in Milton and Wootton, alongside the same line; (e) Additional lands in Harlestone, alongside the same line.

217. London & North Western Railway. Francis Stevenson, engineer. Deposited 29 Nov. 1893. Act 57 & 58 Vict. c. xcii (1894).

In Northants.: Additional lands in East Haddon, Brington and Long Buckby, alongside the Northampton to Rugby line, near the road from Great Brington to Long Buckby.

218. Manchester, Sheffield & Lincolnshire Railway. Deviation Railways Etc. Sir Douglas & Francis Fox, Edward Parry, engineers. Deposited 30 Nov. 1893. Act 57 & 58 Vict. c. lxxxi (1894).

In Northants.:
Deviation Railway No 2. From a junction in Shawell (Leics.) with Railway No 2 authorised by the Extension to London Act, 1893, through Churchover, Newton & Biggin, Brownsover, Clifton on Dunsmore, Rugby, Hillmorton (Warwicks.); Barby (Northants.); Willoughby (Warwicks.); terminating in a junction with Railway No 3 authorised by the 1893 Act (see **212**).

Deviation Railway No 3. From a junction with Deviation Railway No 2 in Clifton on Dunsmore, through Newton & Biggin and Brownsover, terminating in Clifton on Dunsmore (Warwicks.) in a junction with the Rugby and Peterborough branch of the LNWR near Clifton Mill Station.

Also plans showing alterations of levels in Woodford cum Membris, Eydon and Sulgrave (Northants.), and the substitution of an embankment for a viaduct in Shuckburgh (Warwicks.), Catesby, Brackley (Northants.) and Turweston (Bucks.).

219. Great Northern Railway. Richard Johnson, engineer. Sir John Fowler Bt, consulting engineer. Deposited 30 Nov. 1894. Act 58 & 59 Vict. c. xxxvi (1895).

In Northants.: (a) Railway at Peterborough, from a junction with the company's main line near the bridge carrying London Road over the line, terminating in a junction with the company's sidings at the northern end of the viaduct carring the main line over the river Nene; (b) Railway at Werrington Junction, from a junction with the company's up goods and coal line near Werrington Junction Signal Box, terminating in a junction with the up line of the Werrington Junction to Helpston Widening, authorised by the GNR Act of 1891 (see **208**) and then under construction, all in Werrington; (c) Widening of the company's main line from near Helpston Station (planned) in Helpston, through Bainton and Maxey (Northants.), and Tallington (station planned), Uffington, Greatford and Braceborough (Lincs.), terminating in Essendine (Rutland) in a junction with the main line near Essendine Station (planned); (d) Additional land at New England, alongside the main line, in Peterborough.

220. Midland Railway. J.A. McDonald, engineer. Deposited 28 Nov. 1894. Act 58 & 59 Vict. c. cxxxiii (1895).

In Northants.:
Cransley Branch Extension. From a junction with the company's Cransley branch in Cransley, terminating in Mawsley near the southern end of the tramway of the Loddington Ironstone Co. Ltd. Plan shows adjoining ironstone workings and tramway.

221. Manchester, Sheffield & Lincolnshire Railway. Sir Douglas & Francis Fox, engineers (Alexander Ross and Edward Parry printed on the cover sheet but crossed out in ink). Deposited 29 Nov. 1894. Act 58 & 59 Vict. c. cxlviii (1895).

In Northants.: Diversion of Brackley Road in Brackley, at

the junction of roads to Brackley from Helmdon, Turweston and Syresham. Sheet also contains a plan of a deviation of the London Extension line near Finchley Road Station in Hampstead (Middlesex).

222. Kettering Electric Lighting. Application by Kettering Urban District Council for a provisional order under the Electric Lighting Acts, 1882 and 1888. With a 1:10,560 Ordnance Survey map showing proposed area of supply. Lodged with the clerk of the peace 29 Nov. 1895. Order confirmed by Act 59 & 60 Vict. c. lxxxiii (1896).

223. Manchester, Sheffield & Lincolnshire Railway. Sir Douglas & Francis Fox, Alexander Ross, Edward Parry, engineers. Deposited 28 Nov. 1895. Act 59 & 60 Vict. c. ccvii (1896).

In Northants.: (a) Additional land in Charwelton and Woodford cum Membris; (b) Diversion of footpath in Helmdon; (c) Diversion of Bridle Road in Helmdon; all alongside the London Extension line.

224. London & North Western Railway. Francis Stevenson, engineer. Deposited 27 Nov. 1896. Act 60 & 61 Vict. c. liii (1897).

In Northants.: (a) Additional lands at Roade near Blisworth Road bridge, on the London and Birmingham railway; (b) Additional lands at Gayton near the Northampton road bridge, also on the London and Birmingham railway; (c) Additional lands in Little Houghton on the Northampton to Peterborough line, near Clifford Hill Mill (planned); (d) Additional lands in Great Oxendon at Clipston & Oxendon Station (planned) on the Northampton to Market Harborough line.

225. Great Northern Railway. Richard Johnson, W.B. Myers-Beswick, engineers. Sir John Fowler Bt, consulting engineer. Deposited 27 Nov. 1896. Act 60 & 61 Vict. c. xl (1897).

In Northants.: New footpath at Lolham Bridges, in Bainton and Helpston, on the main line from London to York.

226. Manchester, Sheffield & Lincolnshire Railway. Sir Douglas & Frank Fox, engineers (Alexander Ross and Edward Parry crossed through). Deposited 27 Nov. 1896. Act 60 & 61 Vict. c. liv (1897).

In Northants.: From a junction with the company's London Extension line (then under construction) in Eydon (Northants.); through Canons Ashby, Moreton Pinkney, Culworth, Thorpe Mandeville (Northants.); Cropredy (Oxon.); Chalcombe (Northants.); Banbury (Oxon.); Warkworth (Northants.); terminating in a junction with the Oxford and Birmingham branch of the Great Western Railway in Warkworth.

227. Peterborough Electric Lighting. Application by Peterborough Electric Light & Power Co. Ltd for a provisional order under the Electric Lighting Acts 1882 and 1888. With 1:10,560 Ordnance Survey map showing intended area of supply. Deposited 30 Nov. 1897. No Act to confirm order.

228. Great Central Railway. Sir Douglas, Francis & Douglas Fox, C.A. Rowlandson, engineers. Deposited 26 Nov. 1897. Act 61 & 62 Vict. c. ccliii (1898).

In Northants.: Additional lands in Woodford cum Membris near the London Extension line and the road from Woodford to Byfield. (Sheet also shows land to be acquired at Newstead (Notts.) alongside the Midland Railway near that company's Newstead Station.)

229. London & North Western Railway. Francis Stevenson, engineer. Deposited 27 Nov. 1897. Act 61 & 62 Vict. c. ccxxxiv (1898).

In Northants.: Additional land in Courteenhall, alongside the Bletchley, Northampton and Rugby railway.

230. Midland Railway. J.A. McDonald, engineer. Deposited 29 Nov. 1898. Act 63 & 63 Vict. c. cvii (1899).

In Northants.: Additional lands at (a) Kettering Station in Kettering; (b) Wellingborough Station in Wellingborough; both stations planned in detail.

231. Nene Valley Waterworks. E.P. Seaton, Walter Beer, engineers. Deposited 21 Nov. 1898. No Act.

Lands in Little Addington and Finedon (Northants.), with service reservoir in latter parish. With four prints of the Nene Valley Water Bill (H.L.) (62 & 63 Vict., 1899), two copies of 'Observations by the chairman of the Public Health Committee [of the county council] and the County Medical Officer of Health' on the bill, and seven letters from various correspondents to C.A. Markham concerning the bill, Feb.–July 1899.

232. Great Northern Railway. Alexander & Ross, engineer. Deposited 30 Nov. 1898. Act 62 & 63 Vict. c. ccii (1899).

In Northants.: (a) Widening at Peterborough Station, with detailed plan of station and adjoining streets, including proposed street diversions; (b) Bridge over river Welland and additional lands at Deeping on the Lincolnshire Loop Line, including lands in Deeping St James (Lincs.) and Newborough (Northants.).

233. Great Central Railway. Sir Douglas, Francis & Douglas Fox, C.A. Rowlandson, Edward A. Parry, engineers. Deposited 29 Nov. 1898. Act 62 & 63 Vict. xcvi (1899).

Sheet of plans of bridle path and footpath diversions in (a) Woodford cum Membris, (b) Catesby and Charwelton (both on the Rugby to London line); (c) Moreton Pinkney, and (d) Culworth (both on the Woodford to Banbury branch, authorised in 1897).

234. Buckingham, Towcester & Metropolitan Junction Railway. R. Elliott-Cooper, engineer. Deposited 30 Nov. 1899. No Act.

Railway No 1. From a junction with the Metropolitan Railway in East Claydon, through Middle Claydon, Steeple Claydon, Padbury and Buckingham, terminating in Radclive (Bucks.).

Railway No 2. From an end-on junction with Railway No 1, through Radclive, Stowe, Biddlesden (Bucks.), Syresham (Northants.), Luffield Abbey (Bucks.), Silverstone, and Whittlebury, terminating in Towcester (Northants.).

Railway No 3. From a junction with Railway No 1 to a junction with the Banbury and Buckingham branch of the LNWR, all in Radclive.

Railway No 4. From a junction with Railway No 2 in Towcester, terminating in Easton Neston (Northants.) in a junction with the Stratford-upon-Avon, Towcester & Midland Junction Railway.

Railway No 5. From a junction with Railway No 2, terminating in a junction with the Northampton & Banbury Junction Railway, all in Towcester.

Railway No 6. From a junction with Railway No 2, terminating in a junction with the Northampton & Banbury Junction Railway, all in Towcester.

235. London & North Western Railway. Francis Stevenson, engineer. Deposited 29 Nov. 1899. Act 63 & 64 Vict. c. ccxv (1900).

In Northants.: Additional lands at Castle Station, Northampton, in the urban district of St James (Northampton), in the triangle formed by the Northampton and Market Harborough line (from Bridge Street Station), Gas Works branch, and river Nene; plan includes southern end of Castle Station.

236. Great Northern Railway. Alexander & Ross, engineer. Deposited 30 Nov. 1899. Act 63 & 64 Vict. c. cxxxix (1900).

In Northants.: (a) Railway and additional lands at New England, Peterborough, the railway (linking two lengths of the up goods line) running between the company's cottages at Lincoln Road and the engine shed (part of which is planned; (b) Additional lands at Werrington, alongside the main line near the village of Werrington.

237. Higham Ferrers Water Co. Thomas Hennell, engineer. Deposited 29 Nov. 1899. Act 63 & 64 Vict. c. xxvii (1900).

Plan of works in Mears Ashby, Sywell and Ecton (Northants.), including a storage reservoir at Old Sywell Mill and a service reservoir near Vicarage Barn, Mears Ashby.

238. Irthlingborough Gas Co. Application by Irthlingborough Gas & Coke Co. Ltd for a provisional order under the Gas and Waterworks Facilities Act 1870, deposited 29 Nov. 1899. Order confirmed by Act 63 & 64 Vict. c. clxxi (1900).

Plans of existing and proposed new works, and a map of intended area of supply.

239. Wellingborough & District Tramroads. Stephen Sellon, engineer. Deposited 30 Nov. 1899. Act 63 & 64 Vict. c. cxxii (1900).

Circuit of tramroads (to be built by the British Electric Traction Co. Ltd or a company incorporated under the intended Act), from Wellingborough Midland Railway Station, through Wellingborough to Finedon, from Finedon to Higham Ferrers (with a branch from Higham Ferrers via Stanwick to Raunds), from Higham Ferrers to Rushden (with several branches within Rushden), from Rushden to Irchester, and from Irchester to Wellingborough via Wellingborough LNWR Station (fuller details in *Gazette* notice).

In Higham Ferrers, Wellingborough, Rushden, Finedon, Raunds, Irchester, Irthlingborough, Chelveston cum Caldecott and Stanwick (Northants.).

Standard gauge.

240. Wellingborough Electric Lighting. Application by the Electrical Power Distribution Co. Ltd for a provisional order under the Electric Lighting Acts 1882 and 1888, and the Electric Lighting Clauses Act 1899, with a 1:10,560 map showing intended area of supply. Deposited 30 Nov. 1899. Order confirmed by Act 63 & 64 Vict. c. clxx (1900).

241. Rothwell (Northampton) Gas Provisional Order. Deposited 28 Nov. 1899.

Application to the Local Government Board by Rothwell Urban District Council for a provisional order empowering them to acquire the Rothwell Gas Light, Coal & Coke Co. Ltd and to carry on the undertaking. With plan of gasworks, showing proposed addition to retort house, and 1:10,560 map showing area of supply (i.e. Rothwell Urban District) and the location of the gas works. Order confirmed by Act 63 & 64 Vict. c. lvi (1900).

242. County Borough of Northampton. Northampton Corporation Tramways. W.D. Gibbins, Borough Engineer. Application by the Corporation for a provisional order. Deposited 30 Nov. 1900. Order confirmed by Act 1 Edw. VII c. cclxxvii (1901).

Tramway No 1. From Kettering Road, near the junction of Kingsley Road and Abington Grove, in St Giles, by a junction with the existing tramway of the Northampton Street Tramway Co., along Kettering Road, terminating in that road in Kingsthorpe, near the entrance to St Matthew's church. Single line.

Tramway No 2. From Wellingborough Road, near its junction with Roseholme Road, in Abington, by a junction with the existing tramway of the company, thence along Wellingborough Road, terminating in that road near Wantage Road. Single line.

243. Kettering Urban District Water. James Mansergh, engineer. Deposited 30 Nov. 1900. Act 1 Edw. VII c. lxxxv (1901).

Plan of intended Orton Reservoir, Malsor Reservoir, and pipes leading from the reservoirs to the existing Warren Hill Service Reservoir and the Clover Hill Pumping Station and Reservoir. In Cransley, Thorpe Malsor, Loddington, Orton, Rothwell and Kettering (Northants.).

244. Wellingborough Electric Lighting. Application by Wellingborough Urban District Council for a provisional order under the Electric Lighting Acts 1882 and 1888, with a 1:10,560 map showing intended area of supply. Deposited 29 Nov. 1900. Order confirmed by Act 1 Edw. VII c. cxxxviii (1901).

245. London & North Western Railway. Francis Stevenson, engineer. Deposited 27 Nov. 1900. No Act (see **246**).

In Northants.: (a) Additional lands in Nether Heyford on London to Birmingham line (plan includes Stowe Ironworks); (b) New bridge and alteration of road levels in Milton on Bletchley, Northampton and Rugby line, near Glebe Farm, Milton; (c) Additional land in Arthingworth at Kelmarsh Station (planned) on the Northampton and Market Harborough line.

246. London & North Western Railway. Francis Stevenson, engineer. Deposited 27 Nov. 1901. Act 2 Edw. VII c. clxix (1902).

In Northants.: Same three schemes as **245.**

247. Midland Railway. J.A. McDonald, engineer. Deposited 28 Nov. 1901. Act 2 Edw. VII c. cli (1902).

In Northants.: Additional lands at Cranford in Barton Seagrave. Plan shows Butlins Sidings on Kettering to Thrapston line, with two ironstone tramway interchanges.

248. Finedon Urban District Water. Mosley & Scrivener, engineers. Deposited 29 Nov. 1901. Act 2 Edw. VII c. v (1902).

Plan shows intended well, pumping station and service reservoir, on Wellingborough Road, and pipeline along Wellingborough Road and Irthlingborough Road, ending at at an intended water tower on Irthlingborough Road.

249. Higham Ferrers & Rushden Water Board. Reginald E. Middleton, engineer. Deposited 28 Nov. 1901. Act 2 Edw. VII c. xii (1902).

Plan shows pipes running from junction with pipes No 2 authorised by Higham Ferrers Water Act (1900; plan **237**) in

Mears Ashby (near Sywell Mill in Sywell), through Earls Barton, Great Doddington (crossing river Nene), Wollaston (Northants.); Podington, Wymington (Beds.), terminating in a service reservoir in Rushden (Northants.).

250. Wellingborough Gas. Application by Wellingborough Gas Light Co. Ltd for provisional order under the Gas and Waterworks Facilities Act 1870 for powers to acquire additional lands and to construct and maintain additional works, etc. No engineer named. Deposited 29 Nov. 1901. Order confirmed by Act 2 Edw. VII c. ciii (1902).

With detailed plan of proposed new gasworks alongside Midland Railway between Union Road and river Nene, and 1:10,560 map showing company's area of supply.

251. Great Northern Railway. Alexander Ross, engineer. Deposited 28 Nov. 1902. Act 3 Edw. VII c. cxxv (1903).

In Northants.: Additional lands at Peterborough, comprising 1–3 Ashwell Cottages, Walpole Street, to the east of the company's line; plan also shows Vine Cottages, Northolme Terrace and adjoining houses (named) on Walpole Street.

252. Midland Railway. J.A. McDonald, engineer. Deposited 29 Nov. 1902. Act 3 Edw. VII c. xxxii (1903).

In Northants.: Additional lands at Peterborough to west of Midland line near Great Northern station; plans show part of station, also MR engine shed and wagon works.

253. Wellingborough & District Tramroads. Stephen Sellon, engineer. Deposited 28 Nov. 1902. No Act.

Application by British Electric Traction Ltd to build a tramroad from junction with Tramroad No 16a authorised by Wellingborough & District Tramroads Act (1900) (see **239**) in the main road between Higham Ferrers and Irthlingborough, near the level crossing at Higham Ferrers LNWR Station, in the parish of Chelveston cum Caldecott; proceeding along the main road; terminating in High Street East, Irthlingborough, in a junction with Tramroad No 18 authorised by 1900 Act near the corner of Lime Street. Single line with passing places.

254. Midland Railway. J.A. McDonald, engineer. Deposited 27 Nov. 1903. Act 4 Edw. VII c. liii (1904).

In Northants.: (a) Additional lands at Kettering Station (detailed plan); (b) Additional lands in Walton, on the Peterborough to Lynn line near bridge over Lincoln Road.

255. Great Northern Railway. Alexander Ross, engineer. Deposited 30 Nov. 1903. Act 4 Edw. VII c. lxxviii (1904).

In Northants.: (a) Additional lands in Peterborough, comprising The Crescent, Thorpe Road, to the west of the GNR and Midland lines at Peterborough Station; (b) Additional lands at Peakirk, north of Peakirk Station (partly planned) on the Lincolnshire Loop Line (London to Boston).

256. Kettering Improvement. Widenings Grange Road. By Kettering Urban District Council. Thomas R. Smith, engineer. Deposited 30 Nov. 1903. Act 4 Edw. VII c. xxiii (1904).

Plan shows Grange Road, between Union Street and Upper Field Street, with intended widening at each end.

257. Finedon Gas. Application by Finedon Gas Co. Ltd for provisional order under Gas and Waterworks Facilities Act 1870. G.F. Bearn, surveyor. Deposited 28 Nov. 1903. Order confirmed by Act 4 Edw. VII c. clxiv (1904).

With 1:96 plan of existing gasworks on Orchard Road, Finedon, showing use of individual buildings and plant, and 1:63,360 map showing position of works and area of supply.

258. Northampton Electric Lighting. Application by Northampton Electric Light & Power Co. Ltd for provisional order to supply electricity in the county borough of Northampton and in Moulton, Moulton Park, Boughton, Church Brampton, Chapel Brampton, Hardingstone, Wootton, Great Houghton, Little Houghton, Milton, Collingtree, Rothersthorpe, Weston Favell, Great Billing, Little Billing, Dallington, Duston, Upton and Kislingbury (Northants.). Deposited 28 Nov. 1903. Order confirmed by Act 4 Edw. VII c. clxxviii (1904).

With 1:10,560 map showing area of supply.

259. Rushden, Higham Ferrers, Irthlingborough, and Thrapston (Rural) Electric Lighting. Deposited 29 Nov. 1904. Application for a provisional order (by whom not stated; no *Gazette* notice attached to plan). No Act.

Wrapper contains only 1:10,560 map showing proposed area of supply.

260. Wellingborough & District Tramroads and Electricity Supply. E.A. Evans, engineer. Deposited 29 Nov. 1904. No Act.

Application to Parliament by British Electric Traction Ltd to extend the time limit in the Wellingborough & District Tramroads Act 1900 (plan **239**) for the construction of lines authorised by that Act, to abandon other lines authorised by the Act, and to build two new lines, viz. (a) a short line near Wellingborough LNWR Station in Irchester (plan includes part of station and adjoining premises); (b) a spur to create a triangular junction between two previously authorised tramroads near Higham Mill in Higham Ferrers (plan includes the mill and adjoining roads in Higham Ferrers, Chelveston cum Caldecott and Stanwick).

Two other sheets of plans show small areas of additional land to be acquired in Wellingborough (near gasworks on London Road), Rushden (either side of Washbrook Road), Finedon (at junction of Wellingborough Road and Harrowden Road), Raunds (near Grove Street), Burton Latimer (two parcels, one on High Street, the other on Church Street), Earls Barton (on Station Road), Irthlingborough (on High Street), Bozeat (near St Mary's church), Irchester (on School Lane), Wollaston (junction of Back Way and Queen's Road), and Thrapston (on Denford Road, near gasworks).

261. Great Western Railway (Additional Powers). W.W. Grierson, engineer. James C. Inglis, consulting engineer. Deposited 29 Nov. 1904. Act 5 Edw. VII c. cxxxix (1905).

In Northants.: Kings Sutton Loop, completing a triangular junction between the company's Oxford to Birmingham and Banbury to Cheltenham lines, in Adderbury (Oxon.) and Kings Sutton (Northants.).

262. Great Western Railway (New Railways). W.W. Grierson, Edward Parry, engineers. James C. Inglis, consulting engineer. Deposited 29 Nov. 1904. Act 5 Edw. VII c. xcviii (1905).

Railway No 1. From a junction in Aynho (Northants.) with the company's Oxford and Birmingham Railway, through Souldern, Fritwell, Somerton, Ardley, Bucknell, Bicester Market End, Launton, Blackthorn, Piddington (Oxon.); Ludgershall, Brill, Dorton, Wotton Underwood, Ashendon (Bucks.); terminating at Ashendon in a junction with the railway (in course of construction) authorised by the Great Western and Great Central Railway Companies Act 1899.

Railway No 2. From a junction with the company's main line in Ealing, terminating in Acton (Middlesex) in a junction with the company's Acton and Northolt railway.

Railway No 3. From a junction with the Acton and Northolt railway in Acton, terminating in Hammersmith (Middlesex) on the northern side of Uxbridge Road.

Railway No 4. From a junction with Railway No 3 to a junction with the West London Railway, all in Hammersmith.

Railway No 5. From Burnham; through Farnham Royal, Stoke Poges, Wexham, Fulmer, Langley Marish, Denham, Iver (Bucks.); Hillingdon, Uxbridge (Middlesex), terminating in Hillingdon East in a junction with Railway No 4 (in course of construction) authorised by the Great Western Railway Act 1899.

Railway No 6. From a junction with the company's Severn Valley railway in Eardington (Salop); terminating in Bushbury (Staffs.) in a junction with the company's Shrewsbury and Birmingham railway.

Railway No 7. From a junction with the Severn Valley railway; terminating in a junction with Railway No 6; all in Eardington.

Railway No 8. From a junction with Railway No 6; terminating in a junction with the Shrewsbury and Birmingham railway; all in Wrottesley (Staffs.).

Railway No 9. From a junction with Railway No 6 in Trysull & Seisdon; terminating in Kingswinford (Staffs.) in a junction with the company's Kingswinford branch near its terminus.

Railway Nos 6–9 to pass through Eardington, Claverley, Quatford, Quatt Jarvis, Bridgnorth (Salop); Bobbington, Trysull & Seisdon, Wombourn, Himley, Wrottesley, Tettenhall, Wolverhampton, Kingswinford (Staffs.).

Plan and book of reference deposited in Northants. only includes Railway No 1.

263. Midland Railway. J.A. McDonald, engineer. Deposited 28 Nov. 1904. Act 5 Edw. VII c. cliii (1905).

In Northants.: Additional lands at Peterborough, to the west of the Midland line on Westwood Road, near premises of Werner Pfleiderer & Perkins Ltd (Engineers) (planned); MR engine shed near Great Northern Station also planned.

264. Higham Ferrers, Rushden and Wellingborough (Rural District) Electric Lighting. Application by Northampton Electric Power & Traction Co. Ltd for a provisional order to supply electricity within Higham Ferrers municipal borough, Rushden urban district, and Wellingborough rural district. Deposited 29 Nov. 1905. Order confirmed by Act 6 Edw. VII c. cx (1906).

With a set of 1:10,560 maps showing intended area of supply.

265. Peterborough Gas. Corbet Woodall & Son, engineers. Deposited 25 Nov. 1905. Act 6 Edw. VII c. lxxxix (1906).

Plan of Peterborough Gas Co. works on St John's Street, showing additional land to be acquired adjoining works.

266. Raunds Gas. Application by Raunds Gas Light & Coke Co. Ltd for a provisional order. Deposited 29 Nov. 1905. Order confirmed by Act 6 Edw. VII c. cxxxiii (1906).

1:10,560 map showing area of supply.

267. Wollaston Gas. Shearman & Archer, architects and surveyors. Application by the Wollaston Gas Coal & Coke Co. Ltd for provisional order. Deposited 29 Nov. 1905. Order confirmed by Act 6 Edw. VII c. cxxxi (1906).

1:63,360 map showing present and intended areas of supply, and 1:250 plan of gasworks with intended new plant and buildings.

268. Great Northern Railway. Alexander Ross, engineer. Deposited 29 Nov. 1905. Act 6 Edw. VII c. cxliv (1906).

In Northants: Additional lands at Lolham Bridges on the company's main London to York line, in Helpston, Bainton and Maxey.

269. Great Northern Railway. Alexander Ross, engineer. 'Informal deposit made 14th January 1905' written on wrapper, no memorial on plan. Act 6 Edw. VII c. cxliv (1906).

In Northants.: Additional lands at Fletton, parish of Fletton Urban, adjoining the company's main line on London Road.

270. Great Western Railway (General Powers). W.W. Grierson, engineer. James C. Inglis, consulting engineer. Deposited 28 Nov. 1908. Act 9 Edw. VII c. lxxxiv (1909).

In Northants.: (a) Aynho Loop. Connecting the company's Oxford and Birmingham railway north of Aynho Station with Railway No 1 authorised under the Great Western Railway (New Railways) Act 1905 (under construction), all in Aynho parish; (b) Diversion of footpath in Aynho on the railway authorised under 1905 Act. Both plans include simple plan of Aynho Station.

271. Northampton Corporation Tramways Extension. Application by the Corporation for a provisional order. Alfred Fidler, engineer. Deposited 28 Nov. 1908. No Act to confirm order.

From a junction with an existing tramway in the Drapery opposite the main gates of All Saints church, along the Drapery and Bridge Street, terminating near the level crossing of the LNWR in Bridge Street, in the parishes of All Saints and Far Cotton. Single line with passing places. Large-scale plan showing frontages to the Drapery and Bridge Street (some occupiers named).

272. Stratford-upon-Avon & Midland Junction Railway. Russell Willmott, engineer. Deposited 29 Nov. 1909. Act 10 Edw. VII & 1 Geo. V c. viii (1910).

In Northants.: Widening in Towcester between Green's Norton Junction (between the SMJR and Northampton & Banbury Junction Railway) west of the station and the bridge over Watling Street near the station (which is planned); also the acquisition of small piece of land east of the station near the junction between the SMJR and the N&BJR.

Another sheet of plans shows additional lands to be acquired at Stratford on Avon Station in Old Stratford & Drayton; at Ettington Station in Eatington parish; and at Kineton Station in Kineton (Warwicks.); all three stations planned and also locomotive depot at Stratford.

273. Burton Latimer Water. Everard, Son & Pick, engineers. Plan (only) deposited 21 Nov. 1907 by Kettering Rural District Council under Waterworks Clauses Act 1847. No Act.

Plan consists of tracings from 1:2500 Ordnance Survey showing course of mains from an existing pumping station in Weekley, through Kettering, Warkton (service reservoir) and Barton Seagrave to Burton Latimer.

274. Northampton Corporation. Tramways and Street Widenings. Alfred Fidler, engineer. Deposited 29 Nov. 1910. Act 1 & 2 Geo. V c. lxiv (1911).

Tramway No 1. From a junction with existing tramway in the Drapery outside All Saints' church, along Bridge Street, across South Bridge, level crossing over LNWR in Bridge Street, along St Leonard's Road and Towcester Road, terminating there at the borough boundary. Plan on enlarged scale, with detail of street frontages, also showing intended street widenings.

Tramway Nos 2 and 5. Forming a triangular junction linking two existing tramways at the junction of Abington Avenue and Wellingborough Road.

Tramway No 3. Continuing an existing tramway along Harborough Road, Kingsthorpe, terminating at the borough boundary.

Tramway Nos 4, 4a, 4b, 4c. From a triangular junction with an existing tramway on Harborough Road at the junction with Kingsthorpe Grove, along Kingsthorpe Grove, Gipsy Lane, Kingsley Road, over a junction with an existing tramway at the junction with Kettering Road, continuing along Abington Grove, Abington Avenue and Park Avenue, terminating in a junction with intended Tramway No 2 in Wellingborough Road near the junction with Park Avenue.

Tramway No 6. Continuing an existing tramway along Kettering Road from a point near St Matthew's church, terminating at the borough boundary on Kettering Road.

Tramway No 7. Continuing an existing tramway along Weedon Road to the borough boundary near the Red House public house.

275. Great Northern Railway. Alexander Ross, engineer. Deposited 28 Nov. 1910. Act 1 & 2 Geo. V c. lxxix (1911).

In Northants.: Widenings on the main line at Peterborough. (a) South of the Great Northern Station in Fletton (Hunts.) and Peterborough (Northants.); (b) at the station itself, with associated road diversions. Detailed plans of station, goods yard, engine shed etc., and adjoining streets.

Also plans of (a) Additional lands at Biggleswade (Beds.); (b) New England, in Walton and Peterborough, adjoining Gilstrap, Earp & Co.'s malthouse; (c) at Little Bytham Station (planned) in Careby and Little Bytham (Lincs.).

276. Midland Railway (Midland and Great Northern Railways Joint Committee). W.B. Worthington, Alexander Ross, engineers. Deposited 30 Nov. 1910. Act 1 & 2 Geo. V c. c (1911).

In Northants.: Additional lands at Dogsthorpe, in Paston and Eye, on the line from Peterborough to Sutton Bridge, adjoining the Star Pressed Brick Co.'s works (outline plan).

Sheet also shows plan of additional lands at Thursford in Thursford and Briningham (Norfolk), on the line from Lynn to Melton Constable.

277. Rushden & District Electric Lighting. Application by Francis Hugh Thornton (Kingsthorpe Hall, Northampton), Brook Sampson (Northampton) and John Clark (Heatherbreea House, Rushden) for a provisional order to supply electricity in Rushden urban district, Higham Ferrers borough, Irthlingborough urban district, Wellingborough rural district, and Chelveston cum Caldecott parish. G.H. Jackson, engineer. Deposited 29 Nov. 1911. Order confirmed by Act 2 & 3 Geo. V c. cxvi (1912).

With a set of 1:10,560 maps showing intended area of supply, and 1:63,360 index sheet.

278. Great Northern Railway. Charles J. Brown, engineer. Deposited 28 Nov. 1912. Act 3 & 4 Geo. V c. lx (1913).

In Northants.: Additional lands at Walton, alongside the company's main line near Sage & Co.'s works. Plan sheet also shows additional lands at Tallington (Lincs.) alongside the main line at Tallington Station (detailed plan).

279. Great Northern Railway. Charles J. Brown, engineer. Deposited 29 Nov. 1912. Act 3 & 4 Geo. V c. lx (1913).

In Northants.: (a) New footpath at Peterborough between Priestgate and Cowgate, near goods depot; (b) Additional lands at Peterborough on the Peterborough to Lynn branch near Lincoln Road bridge. Sheet of plans also includes additional lands at (a) Lound (Lincs.) on the Midland & Great Northern Joint Railway line from Saxby to Bourne; (b) Twenty Station, Bourne, on the M&GNJR line from Bourne to Spalding; (c) Holt (Norfolk), near Holt Station on the M&GNJR line from Melton Constable to Cromer.

280. Wellingborough Electric Lighting. Application by Wellingborough Electric Supply Co. Ltd for a provisional order. Deposited 30 Nov. 1912. Order confirmed by Act 3 & 4 Geo. V c. cliii (1913).

Deposit comprises plan of company's existing yard at Cannon Street, Wellingborough; 1:2500 map showing premises within a radius of 300 yards of the yard; and a book of reference (a) for land to be taken compulsorily to extend the yard, and (b) of owners and lessees within radius of 300 yards.

281. Northampton Corporation Water. Frank Tomlinson, engineer. Deposited 29 Nov. 1912. Act 3 & 4 Geo. V c. xv (1913).

Application to Parliament by the Corporation to extend the limits of supply to include Guilsborough, Coton and Hollowell; recites Northampton Corporation Waterworks Acts 1861–84.

Work No 1. Impounding reservoir (to be called the Hollowell Reservoir) in Guilsborough and Hollowell, formed by a dam across the valley of the Stowe Brook.

Work No 2. An aqueduct from the dam in Hollowell, terminating in Ravensthorpe, at the Corporation's pure water tank.

Work No 3. An aqueduct from a junction with the existing pipes near the Corporation's engine-house in Ravensthorpe; through Teeton, Holdenby, Spratton, Chapel Brampton, Boughton, Moulton Park and Kingsthorpe; terminating in Boughton at the Corporation's Boughton Service Reservoir.

Work No 4. A service reservoir (to be called Boughton Service Reservoir) in Boughton adjoining the existing service reservoir.

Work No 5. An aqueduct from the existing sandwell of the Corporation in Ravensthorpe, terminating in Hollowell at Stowe Brook.

With a notice, dated 17 May 1918, addressed by the Town Clerk of Northampton to the Clerk of the County Council, under the Northampton Corporation Act 1911, the Northampton Corporation Act 1913 (*sic; recte* Northampton Corporation Water Act 1913, as above) and the Special Acts (Extension of Time) Act 1915, stating that the Corporation have applied to the Local Government Board for an order to extend to 18 Aug. 1919 the time limit under the 1911 Act (as extended by an order of 7 July 1917) for the widening of Bridge Street, and also for an order to extend to 4 July 1919 the time limit under the 1913 Act (as extended in 1917) for the compulsory purchase of lands under the 1913 Act.

282. [Northampton Corporation Water]. Wrapper contains three sheets of plans numbered 4, 5 and 6 (i.e. from the scheme of 1913, above **281**), to which is attached a letter dated 21 May 1919 from the Borough Water Engineer to the County Surveyor, stating that the plans show the authorised water main from Ravensthorpe to Northampton which the Corporation is proposing to lay at an early date and which will, as far as possible, be laid under roadside waste.

283. Northampton Corporation (Water). Frank Tomlinson, engineer. Deposited 29 Nov. 1921. Act 12 & 13 Geo. V c. xxv (1922).

Plans of works in Chapel Brampton, Pitsford and Broughton (Northants.).

284. Midland Railway. James Briggs, engineer. Deposited 29 Nov. 1922. Act 13 & 14 Geo. V c. lxxxii (1923).

In Northants.: (a) Additional lands at Corby, south of Corby & Weldon Station (partly planned), on the Kettering to Manton line; (b) Additional lands at Desborough, west of Desborough & Rothwell Station, on the Leicester to Bedford line; (c) Additional lands at Market Harborough in Braybrooke, near Braybrooke Lodge, also on the Leicester to Bedford line.

285. Rugby Urban District Council. General Powers Bill. G. Bertram Kershaw, engineer. Deposited 30 Nov. 1922 under Waterworks Clauses Act 1847. Act 13 & 14 Geo. V c. lxxv (1923).

With correspondence between the clerk of Rugby UDC and the Clerk of the Peace for Northants. concerning correction of errors found in the plans and book of reference, June–July 1926.

Works in connection with new waterworks in South Kilworth, Westrill & Starmore (Leics.); Stanford, Welford (Northants.); Brownsover (Warwicks.); Swinford, Catthorpe (Leics.); Lilbourne (Northants.); Newton & Biggin, Clifton on Dunsmore (Warwicks.); Husbands Bosworth (Leics.); Rugby (Warwicks.).

286. Northampton Electricity Extension. Deposited 10 Oct. 1922. No Act.

Wrapper contains only a 1:10,560 map showing proposed extension of area of supply, in Overstone, Sywell, Mears Ashby, Ecton and Earls Barton (Northants.).

287. Rushden & District Electricity Extension. Application by Rushden & District Electric Supply Co. Ltd to the Electricity Commissioners for a Special Order under the Electricity Supply Acts 1882–1922 to extend the area of supply under the Rushden & District Electric Lighting Order 1912 (see **277**) so as to include Raunds urban district, and the parishes of Stanwick, Chelveston cum Caldecott (except that part which lies to the west of the road from Higham Ferrers to Chelveston Lodge), Newton Bromswold, Higham Ferrers, Irchester (that part which lies east of the LMS Railway) (Northants.); Wymington, Podington, Harrold, Odell, Souldrop, Sharnbrook, Bletsoe, Riseley, Knotting, Melchbourne and Yelden (Beds.). Dated 11 June 1923. No endorsement or stamp concerning deposit.

Deposit comprises *Gazette* notice, print of draft special order, and 1:10,560 map showing intended area of supply.

288. Wellingborough Electricity Extension. Application by Wellingborough Electric Supply Co. Ltd to the Electricity Commissioners for a special order under the Electricity (Supply) Acts 1882–1922 to extend the area of supply under the Wellingborough Electric Lighting Order 1900 and 1913 (see **244**, **280**) to include Burton Latimer, Finedon, so much of Irthlingborough as lies west of the road leading from the Wellingborough and Higham Ferrers road through Ditchford LMS Station to the Wellingborough and Irthlingborough road, and Isham, Orlingbury, Little Harrowden, Great Harrowden, Hardwick, Wilby, Great Doddington, Wollaston, Strixton, Grendon, Easton Maudit, Bozeat and Irchester (except that part of Irchester which lies east of the LMS main line). Dated 11 June 1923. No endorsement or stamp concerning deposit.

Deposit comprises *Gazette* notice, print of draft special order, and 1:10,560 map showing intended area of supply.

289. Kettering Electricity Extension. Application by Kettering Urban District Council to the Electricity Commissioners for a special order under the Electricity Supply Acts 1882–1922 to extend the area of supply under the Kettering Electric Lighting Order 1896 (see **222**) to include Market Harborough (Leics.); Desborough and Rothwell urban districts; and Lubenham (Leics.); Brampton Ash, Braybrooke, Dingley, East Farndon, Barton Seagrave, Burton Latimer, Broughton, Cransley, Geddington, Loddington, Orton, Pytchley, Rushton, Thorpe Malsor, Warkton, Weekley and Isham (Northants.). Dated 14 June 1923. Stamped receipt by Clerk of the County Council dated 15 June 1923.

Deposit comprises *Gazette* notice, draft print of special order, and 1:10,560 map showing intended area of supply.

290. Northampton Electricity (Extension). Application by the Northampton Electric Light & Power Co. Ltd to the Electricity Commissioners for a special order under the Electricity (Supply) Acts 1882–1922 to extend the area of supply under the Northampton & District Electricity Orders 1890–1922 (see **200**, **206**, **258**).

The extension includes Hannington, Holcot, Walgrave, Old, Scaldwell, Brixworth, Pitsford, Hanging Houghton, Lamport, Cottesbrooke, Great Creaton, Spratton, Holdenby, Teeton, Hollowell, Guilsborough, Coton, Ravensthorpe, East Haddon, Brington, Althorp, Harlestone, Harpole, Upper Heyford, Nether Heyford, Bugbrooke, West Haddon, Watford, Long Buckby, Welton, Whilton, Norton, Staverton, Badby, Newnham, Everdon, Dodford, Weedon Beck, Brockhall, Flore and Stowe Nine Churches, Daventry, Cold Higham, Green's Norton, Towcester, Whittlebury, Pattishall, Gayton, Blisworth, Tiffield, Easton Neston, Shutlanger, Stoke Bruerne, Potterspury rural district, Courteenhall, Roade, Quinton, Preston Deanery, Piddington, Horton, Hackleton, Brafield on the Green, Cogenhoe, Whiston, Denton, Castle Ashby, Yardley Hastings (Northants.); Wolverton, Newport Pagnell, Newport Pagnell rural district, and Bletchley (Bucks.). Dated 24 Jan. 1924; endorsed with stamped receipt of even date.

Deposit comprises *Gazette* notice, draft print of special order, and 1:10,560 map showing intended area of supply.

291. Rushden & District Electricity (Extension). G.H. Jackson, engineer. Application by the Rushden & District Electric Supply Co. Ltd to the Electricity Commissioners for a special order to extend the area of supply under the Rushden & District Electricity Orders 1912 and 1923 (see **277**, **287**) to include Oundle, Hargrave, Great Addington, Little Addington, Ringstead, Titchmarsh, Clopton, Thrapston, Islip, Twywell, Woodford, Slipton, Lowick, Aldwincle, Sudborough, Brigstock, Benefield, Glapthorn, Ashton [near Oundle], Lutton, Polebrook, Armston, Hemington, Luddington, Thurning, Barnwell All Saints, Barnwell St Andrews, Lilford cum Wigsthorpe, Thorpe Achurch, Wadenhoe, Pilton, and Stoke Doyle (Northants.). Dated 19 Dec. 1924. Stamped receipt dated 17 Dec. 1924.

Deposit comprises *Gazette* notice, draft print of special

order, and 1:10,560 map showing intended area of supply.

292. Northampton Gas. Draft special order proposed to be made by the Board of Trade under the Gas Regulation Act 1920 on the application of the Northampton Gaslight Co. (to be known collectively with earlier provisions as the Northampton Gas Acts and Orders 1858–1925). With stamped receipt dated 21 Jan. 1925 ('Deposited' pencilled against stamp).

Includes an extension of the area of supply to include Great Creaton, Hanging Houghton, Hannington, Holcot, Lamport, Scaldwell, Walgrave, Flore, Weedon Beck, Bugbrooke, Nether Heyford, Upper Heyford, Easton Neston with Hulcote, Gayton, Tiffield, Brafield on the Green, Cogenhoe, Hackleton, Horton, Piddington, such parts of Preston Deanery and Quinton as are not already included in the limits, and Sywell (Northants.).

With a 1:63,360 map showing intended area of supply.

293. Brackley Electricity. 1:10,560 map showing intended area of supply under a proposed Brackley Electricity Special Order. Stamped receipt dated 17 March 1927.

With a file of correspondence between the clerk of the county council, the undertakers, and the Electricity Commissioner concerning an application by Cecil James Cooper and William Cecil Hanis, both of 14 Gloucester Road, South Kensington, London SW7, for an order to supply electricity within the borough of Brackley, terminating in notification by the Commission to the clerk that the undertakers have withdrawn their application for an order and their opposition to an application for a special order by the Northampton Electric Light & Power Co. Ltd to supply (inter alia) the borough of Brackley (see **294**). March 1927–Oct. 1928.

294. Northampton Electricity (Extension). G.H. Jackson, engineer. Set of 1:10,560 maps showing intended area of supply.

With a file of correspondence between the clerk of the county council and local authorities, the undertakers, the Electricity Commission and others, with related papers. March 1927–Oct. 1928. Terminates in prints of:

(a) *Northampton Electricity (Extension) Special Order 1928.* Known collectively with the Northampton Electric Lighting Order 1890, the Northampton & District Electric Lighting Order 1904, and the Northampton Electricity (Extension) Special Orders 1923 and 1924 as the Northampton & District Electricity Orders 1890–1928. Includes provision for extending supply to Crick rural district, Cold Ashby, Thornby, Naseby, Haselbech, Maidwell, Draughton, Faxton, Mawsley, Welford and Sulby (Northants.).

(b) *Brackley, Buckingham and Winslow Electricity Special Order 1928.* Area of supply includes Brackley borough, Middleton Cheney and Brackey rural districts, Catesby, Hellidon, Charwelton, Fawsley, Byfield, Woodford cum Membris, Canons Ashby, Preston Capes, Farthingstone, Braunston, Ashby St Ledgers, Winwick, Litchborough, Maidford, Blakesley, Adstone, Woodend, Plumpton, Weedon Lois, Bradden, Slapton, Wappenham, Abthorpe and Silverstone (Northants.); Buckingham borough, and Buckingham and Winslow rural districts (Bucks.).

The undertakers under both orders are the Northampton Electric Light & Power Co. Ltd.

295. Mid-East England Electricity. Set of 1:63,360 maps showing intended area of supply, with an index sheet showing transmission lines, transformers, sub-stations etc. Deposited 26 June 1929.

With two copies of a letter from the Electricity Commission to the clerk of the county council announcing the result of a local inquiry into the proposed Mid-East England, East Leicestershire, Kettering (Extension), and Stamford Rural Electricity Special Orders, 18 Dec. 1930.

296. Kettering Electricity (Extension) Special Order, 1929. Kettering Urban District Council, undertakers. No date of deposit.

Deposit includes 1:63,360 map showing area of supply and print of draft Kettering Electricity (Extension) Special Order 1929, to be cited collectively with the Kettering Electric Lighting Order 1896 and the Kettering Electricity (Extension) Special Order 1923 as the Kettering Electricity Orders 1896–1929 (see **222, 289**).

Area of supply extended to include Beanfield Lawns, Corby, Cottingham, Cranford St Andrew, Cranford St John, East Carlton, Grafton Underwood, Great Oakley, Great Weldon, Harrington, Little Oakley, Little Weldon, Middleton, Newton, Stanion, Gretton, Rockingham, Arthingworth, Ashley, Clipston, Great Oxendon, Hothorpe, Kelmarsh, Marston Trussell, Sibbertoft, Stoke Albany, Sutton Basset, Thorpe Lubenham, Weston by Welland and Wilbarston (Northants.).

Also a file of correspondence between the clerk of the county council and local authorities, the Electricity Commission and others, Nov. 1929 to May 1931. Terminates in print of Kettering Electricity (Extension) Special Order 1931, in which only the orders of 1896 and 1923 are recited (not that of 1929 above), and the added area of supply is the same as in 1929, with the omission of Gretton and Rockingham (i.e. the two parish in Gretton rural district).

297. Stamford Rural Electricity Special Order, 1930. Urban Electric Supply Co. Ltd, undertakers. Edmundson's Electricity Corporation Ltd, engineers. Stamped receipt dated 20 Jan. 1930.

Deposit includes a set of 1:10,560 maps showing area of supply, print of draft special order, and copy correspondence, Jan. to Dec. 1930, relating to the order, with a note that original correspondence is on the file relating to the Kettering Electricity (Extension) Special Order 1929–30 (see **296**).

The area of supply includes Barnack, Stamford Baron, Wothorpe, Easton on the Hill, Collyweston (Northants.); Ketton and Tinwell (Rutland).

298. Northampton Gas Special Order, 1930. Only items in wrapper (deposited 12 Aug. 1930) are:

(a) 1:63,360 map showing existing and new areas of supply and gas works at Northampton and Weedon Beck (belonging to 'Northampton Co.'), and Towcester (belonging to 'Towcester Co.'); the added areas comprise Mears Ashby, Greens Norton, Towcester and Paulerspury (Northants.).

(b) 1:2500 map showing land occupied by the gas works in Towcester.

299. Corby (Northants.) & District Water. Herbert Lapworth, engineer. Deposited 20 Nov. 1930. Act 21 & 22 Geo. V c. civ (1931).

Works in Stoke Dry (Rutland and Leics.); Caldecott (Rutland); Stockerston, Great Easton (Leics.); Rockingham, Corby (Northants.). Wrapper includes (as well as plans and book of reference) two copies of the notice to owners addressed to the county council as highway authority.

300. Grand Union Canal. Sir R. Elliott-Cooper & Son, engineers. Deposited 19 Nov. 1930 (formal memorial, including hour of day, signed by the clerk of the peace (not the clerk of the county council), not a stamp as used for gas and electricity orders). Act 21 & 22 Geo. V c. xc (1931).

Works in Knowle, Hatton, Budbrooke, Warwick, Offchurch, Radford Semele, Ufton, Long Itchington, Stockton, Grandborough, Napton on the Hill, Solihull, Birmingham, Shrewley, Lower Shuckburgh, Wolfhampcote (Warwicks.); Braunston (Northants.); Heston & Isleworth (Middlesex).

301. London & North Eastern Railway. C.J. Brown, J. Miller, engineers. Deposited 18 Nov. 1930 (memorial as in **300**). Act 21 & 22 Geo. V c. xcii (1931).

In Northants.: Proposed diversion of road in Woodford cum Membris, near Woodford & Hinton Station, on the Rugby to London Marylebone line.

302. Grand Union Canal (Leicester Canals Purchase &c.). Sir R. Elliott-Cooper & Son, engineers. Deposited 30 April 1931. Act 21 & 22 Geo. V c. cvii (1931).

Works in Watford (Northants.); Foxton, Gumley (Leics.).

303. Bye-laws as to Petroleum Filling Stations made by the Northamptonshire County Coucil under Section 11 of the Petroleum (Consolidation) Act, 1928. Print sealed 14 Nov. 1931; confirmed by the Secretary of State, 23 Dec. 1931.

With 13 deposited maps.

No 1 is a map of Northamptonshire showing (a) area of the administrative county to which the bye-laws in Part I of the Bye-laws relate (viz. the entire administrative county less the county borough, municipal boroughs and urban districts); (b) the location of war memorials (under Part II of the Bye-laws the erection of a filling station is prohibited within a radius of 100 yards of a war memorial); (c) the location of ancient monuments, with the radius in yards around each within which the erection of a filling station is also prohibited under Part II of the Bye-laws).

Nos 2–13 are a set of 1:10,560 maps of the administrative county showing (a) the boundaries of areas within which the erection of filling stations is regulated under Part I of the Bye-laws (as (a) above); (b) areas in which the erection of filling stations is prohibited under Part II of the Bye-laws.

304. Kettering Gas Special Order, 1934. Geo. Evetts, engineer. Deposited 12 Dec. 1934. With copy letters of even date from the clerk of the peace to Kettering Gas Co., noting that deposit is made under the Gas (Special Orders) Rules 1922.

(a) Set of 1:63,360 maps showing existing and additional areas of supply.

(b) 1:2500 maps showing (i) site of proposed gas works in Corby; (ii) additional land for storage of gas adjoining existing gas works in Kettering.

With a print of draft Kettering Gas Special Order 1934, to be cited with the Kettering Gas Act 1891 (54 & 55 Vict. c. xci), the Kettering Gas Charges (Order) 1921 (S.R. & O. 1921 No 1997), and the Kettering Gas Act 1932 (22 & 23 Geo. V c. lxiii) as the Kettering Gas Acts and Orders 1891–1934.

Undertaker is Kettering Gas Co. Added supply areas comprise Burton Latimer urban district, part of Rothwell urban district (described in First Schedule to Order), Isham, Brigstock, Rockingham, Barford, Barton Seagrave, Beanfield Lawns, Broughton, Corby, Cottingham, Cranford St Andrew, Cranford St John, Cransley, East Carlton, Geddington, Glendon, Grafton Underwood, Harrington, Loddington, Middleton, Newton, Great Oakley, Little Oakley, Orton, Pytchley, Rushton, Stanion, Thorpe Malsor, Warkton, Weekley, Great Weldon and Little Weldon (Northants.).

Also a print of the Kettering Gas Order 1935 (S.R. & O. 1935 No 784). Dated 2 Aug. 1935. Recites the Acts of 1891 and 1932 and the order of 1921, but not the order of 1934 above, as previous legislation, to be cited collectively as the Kettering Gas Acts and Orders 1891–1935.

Undertaker is Kettering Gas Co. Added area of supply includes so much of Kettering urban district as is not within the existing limits of supply; Burton Latimer urban district, and Brigstock, Broughton, Corby, Cottingham, Cranford, Cransley, East Carlton, Geddington, Grafton Underwood, Harrington, Middleton, Newton, Oakley, Orton, Pytchley, Rockingham, Rushton, Stanion, Thorpe Malsor, Warkton, Weekley, Weldon, and so much of Loddington as consituted that parish prior to 1 April 1935 (Northants.).

305. Northampton Gas Special Order, 1933. Contents of wrapper includes a receipt from the clerk of the peace dated 19 Dec. 1932 for the following items received from solicitors to the Northampton Gaslight Co.:

(1) Two prints of draft special order under Gas Undertakings Acts 1920 and 1929.

(2) Print of intended advertisement of same.

(3) 1:63,360 map showing existing gas works at Northampton and Towcester (Northampton Gaslight Co.) and Long Buckby (Long Buckby Gas-Light Coke & Coal Co. Ltd) and boundaries of proposed and existing areas of supply.

(4) 1:2500 map showing land proposed to be used for construction of gas works, viz. the site of existing works at Long Buckby.

Of which the wrapper now contains (3), (4) and one print of the draft special order.

The draft special order recites the Northampton Gas Acts and Order 1858–1931 as previous legislation; names the undertaker as the Northampton Gaslight Co.; provides for the acquisition of the Long Buckby Gas-Light Coke & Coal Co. Ltd (registered 1859) by the Northampton Co. and the winding-up of the Long Buckby Co.; and extends the area of supply to include Pattishall, Brockhall, Dodford, Long Buckby, West Haddon, Whilton, Coton, East Haddon, Guilsborough, Holdenby, Hollowell, Ravensthorpe and Teeton (Northants.).

306. Northampton Gas, 1927. Wrapper contains only (a) 1:63,360 map showing the position of the works of the Northampton Gaslight Co. at Northampton and those of the Weedon Gaslight Coke & Coal Co. at Weedon Beck; (b) 1:2500 map showing the existing gas works at Weedon and what is presumably the proposed area of extension alongside. (b) bears a receipt stamp dated 10 Aug. 1927.

307. Wellingborough Gas, 1935. None of the following bears a date of deposit.

(a) Application by the Wellingborough Gas Light Co. Ltd for a special order to purchase the Irthlingborough Gas & Coke Co. Ltd, take over the limits of supply of the Irthlingborough Co. (viz. Irthlingborough urban district, Little Addington, Great Addington, Woodford, and that part of Chelveston cum Caldecott which was supplied with gas by the Irthlingborough Co. under the Irthlingborough Gas Order 1900 and excluded from the limits of supply of the Rushden & Higham Ferrers District Gas Co. by the Rushden & Higham Ferrers District Gas Act 1899 (62 & 63 Vict. c. 1)), and add to the limits of supply Great Harrowden, Little Harrowden, Orlingbury, Hardwick and Castle Ashby.

With a print of the draft special order, a set of 1:63,360 maps showing existing and new areas of supply and position of Wellingborough and Irthlingborough gas works, and a set of

1:2500 maps showing the land occupied by the works of the two companies.

(b) Application by the Wellingborough Gas Light Co. Ltd for a special order to extend the limit of supply to included so much of the former Irthlingborough urban district as was transferred to the Wellingborough urban district by the County of Northampton (Wellingborough Extension) Review Order 1935. With a print of the draft Special Order (Wellingborough Gas (No 2)) and a 1:63,360 map showing added area of supply.

308. Banbury Gas, 1936. No date of deposit. Application dated 25 Feb. 1936 by Banbury Gaslight & Coke Co. for a special order empowering the company to acquire additional land on which to extend their works in the borough of Banbury. With two prints of the *Gazette* notice and the draft special order; 1:63,360 maps showing existing works, proposed extension, and area of supply (of which latter includes adjoining parishes in Northants.); and a 1:2500 map showing land to be taken for extension.

309. Northampton Gas, 1937. Deposited 7 Oct. 1936.

(a) Print of draft special order.

(b) 1:63,360 map showing existing supply areas of the Northampton and Newport Pagnell gas companies and the location of their respective works.

(c) 1:2500 maps showing land occupied by the works at Olney and Turvey (Bucks.).

With a print of the Northampton Gas Order 1937, dated 11 June 1937 (S.R. & O. 1937 No 532). May be cited with earlier Acts and orders as the Northampton Gas Acts and Orders 1858–1937. Undertaker is the Northampton Gaslight Co. Provides for the transfer of the undertakings of the Newport Pagnell Gas & Coke Co. Ltd (incorporated 1874) and the Olney Gas Light Coke & Coal Co. Ltd (established 1854) to the Northampton Co., and the extension of the area of supply to include the area of the Newport Pagnell Co., together with Ashton [near Roade], Denton, Hartwell, Yardley Hastings, the Whiston Ward of Cogenhoe, Furtho, Grafton Regis, Potterspury, Shutlanger, Stoke Bruerne, Yardley Gobion (Northants.); Clifton Reynes, Cold Brayfield, Emberton, Hanslope, Lavendon, Newton Blossomville, Olney, Ravenstone, Tyringham with Filgrave, Warrington, Weston Underwood (Bucks.); Carlton & Chellington, Harrold, Odell, Turvey (Beds.).

310. Stony Stratford Gas Special Order, 1937. Receipt dated 10 Feb. 1937 for:

(a) Print of draft special order. The undertaker is the Stony Stratford Gas & Coke Co. Ltd; limit of supply includes part of Wolverton urban district (Bucks.) and Cosgrove and Passenham (Northants.).

(b) 1:63,360 map showing limits of supply.

(c) 1:2500 map showing lands in Stony Stratford used for gas purposes.

311. London & North Eastern Railway. R.J.M. Inglis, engineer. Stamped receipt dated 18 Nov. 1937. Act 1 & 2 Geo. VI c. liii (1938).

In Northants.: Additional lands in Charwelton, near Charwelton Station on the London to Sheffield main line.

312. Wellingborough Gas Special Order, 1938. No date of deposit.

(a) Print of draft Wellingborough Gas Order 1938, which may be cited with earlier orders as the Wellingborough Gas Orders 1879–1938. Undertaker is the Wellingborough Gas Light Co. Ltd. Provides for transfer of the Wollaston Gas Coal & Coke Co. Ltd, repeal of the Wollaston Gas Order 1906 (confirmed by the Gas and Water Orders Confirmation Act 1906 (6 Edw. VII c. cxxxi)), and the addition to the area of supply of (i) part of Wollaston, Strixton and Bozeat (formerly the Wollaston limits), and (ii) the remainder of Wollaston, Grendon and Easton Maudit.

(b) 1:63,360 maps showing existing and added areas of supply and the position of the Wellingborough and Wollaston gas works.

(c) 1:2500 map showing the lands occupied by the gas works at Wollaston.

313. United District Gas Order. No date of deposit; wrapper endorsed 1938.

(a) Draft special order made by the Board of Trade under the Gas Undertakings Acts 1920–34 on the application of the United Gas Company. May be cited as the United District Gas Order 1938, and cited with the United Gas Act and Orders 1913–28 as the United Gas Act and Orders 1913–38. Provides for absorption of the gas undertaking in Bampton and Aston Bampton (Oxon.) of the General Gas & Electricity Co. Ltd. Extended supply area includes Chipping Norton borough, Bicester urban district, parishes in Witney, Chipping Norton, Banbury and Ploughley rural districts (Oxon.); Buckingham borough, Buckingham rural district (Bucks.); Brackley borough, Brackley rural district (except Chalcombe, Middleton Cheney, Warkworth and Kings Sutton), Byfield and Woodford cum Membris in Daventry rural district (Northants.); parishes in Cricklade & Wootton Bassett rural district and part of a parish in Highworth rural district (Wilts.).

(b) 1:63,360 map showing supply area, existing authorised gasworks at Woodford (Northants.), Buckingham (Bucks.), Adderbury, Bicester, Charlbury and Shipton under Wychwood (Oxon.), Cricklade (Wilts.), and existing gasworks to be authorised at Bampton (Oxon.).

(c) 1:2500 map showing land to be used for gasworks purposes in Bampton (Oxon.).

314. Kettering Gas Order. No date of deposit; wrapper endorsed 1942.

(a) Set of 1:63,360 maps titled Board of Trade 1940. Kettering Gas (Special Order), showing existing area of supply, area within which Kettering Gas Co. are precluded from supplying without the consent of the Burton Latimer Gas Co. Ltd, the situation of the existing gasworks of the Kettering Gas Co., and the situation of proposed works (i.e. the existing works of the Burton Latimer Gas Co. Ltd).

(b) 1:2500 map showing gasworks of Burton Latimer Gas Co. Ltd near Burton Latimer Station (LMS), in Isham.

315. Stony Stratford Gas Special Order 1943. No date of deposit. No print of draft order; title taken from wrapper.

(a) Two 1:63,360 maps showing existing area of supply in Stony Stratford (Bucks.), Passenham, Deanshanger, Old Stratford and Cosgrove (Northants.), and existing and proposed gasworks at Stony Stratford.

(b) 1:2500 plan showing existing Stony Stratford gasworks and small adjoining area of land proposed to be used for an extension to the works.

316. Northampton Corporation Bill. Plans and sections and book of reference (in duplicate), with notice addressed to the county council as the owner of certain highways affected, for proposed waterworks. Deposited (formal memorial) 18 Nov. 1942. Leonard H. Brown, engineer. T. & C. Hawksley, consulting engineers. Act 6 & 7 Geo. VI c. xv (1943).

Works in Boughton, Brixworth, Holcot, Moulton, Old, Pitsford, Scaldwell and Walgrave (Northants.).

317. British Transport Commission. J. Taylor Thompson, engineer. Deposited (formal memorial) 11 Nov. 1954. Act 4 & 5 Eliz. II c. xxx (1955).

In Northants.: Diversion of footpath at Blisworth (in Gayton) on the Weedon to Blisworth line.

318. Great Ouse Water. Binnie, Deacon & Gourley, consulting engineers. Deposited 18 Nov. 1960. Act 9 & 10 Eliz. c. xlii (1961).

Works in Rushden, Wellingborough, Hannington, Great Doddington, Hardwick, Irchester, Little Harrowden, Newton Bromswold (Northants.).

SOKE OF PETERBOROUGH

DEPOSITED PLANS

A map of the intended canal and navigation from the Union Canal near Harborough in the county of Leicester to Stamford and Spalding in the county of Lincoln and Peterborough in the county of Northampton.
Deposited 28 Sept. 1810. Map 1834.
Northants. Canal Plan 14.

A map of the intended canal from the Union Canal near Harborough in the county of Leicester to the Welland Navigation at Stamford in the county of Lincoln.
Deposited 30 Sept. 1810. Map 1862.
Northants. Canal Plan 13.

Nene Outfall.
Plan only, deposited 29 Oct. 1822. Map 1803.
Northants. Canal Plan 16.

A plan and section of the river from and above the city of Peterborough in the county of Northampton; to and below the town of Wisbech in the Isle of Ely and county of Cambridge. 1823.
Deposited 30 Sept. 1823. Map 1819.
Northants. Canal Plan 17.

Plan and sections of the intended improvement to the outfall of the river Nene by a new cut or channel from Kinderley's Cut to Crabhole.
Deposited 11 Oct. 1826. Map 1822.
Northants. Canal Plan 20.

Plan and section of certain rivers or cuts called Hills Cut and Smith's Leam and of that part of the river Nene extending from Hills Cut to Goldiford Stanch of that part of the Wisbech river extending from Smith's Leam to the limits of the port of Wisbech. 1826.
Deposited 11 Nov. 1826. Map 1863.
Northants. Canal Plan 19.

Plan and section of proposed new drain from Clow's Cross to the new outfall near Buckworth Sluice. 1828.
No plan; book of reference for this scheme (Northants. Canal Plan 22) boxed with Map 1838 (see next entry below).

Plan and section of the intended variation and extension of the Nene Outfall Cut from Sutton-Wash to Buckworth Sluice. 1828.
Deposited 28 Nov. 1828. Map 1838. Boxed with book of reference for the proposed cut from Clow's Cross to Buckworth Sluice, also deposited in 1828 (see previous entry).
Northants. Canal Plan 23.

Northern & Eastern Railway. 1836.
(a) Plan, no memorial. Map 1871.
(b) Plan, lacking cover, deposited 30 Nov. 1836. Map 1873.
Includes sections of Sleaford and Boston branches not in Northants. copy.
Northants. Main Series 6.

Railway commencing by a junction with the London & Birmingham Railway in the parish of Gayton; terminating near Peterborough. 'Gayton' has been altered in MS to Blisworth on the title-page of the plans.
(a) Book of reference, deposited 30 Nov. 1842. Map 1865.
(b) Plan, deposited 30 Nov. 1842. Map 1894.
(c) Plan, deposited 30 Nov. 1842. Map 1827.
Northants. Main Series 10.

River Nene Improvements &c. 1840.
(a) Plan, no memorial. Map 1815.
(b) Plan, no memorial. Map 1906.
Northants. Main Series 10a.

Railway commencing by a junction with the Northern & Eastern Extension Railway, in the parish of Newport (Essex), and terminating at Thetford (Norfolk), also of an intended railway from Ely to Peterborough, and of a branch railway to Wisbeach (Cambs).
(a) Plan, deposited 30 Nov. 1843. Robert Stephenson, engineer. C.F. Cheffins, surveyor.
[Newport to Thetford:] From Newport, through Wendens Ambo, Littlebury (Essex); Ickleton (Cambs.); Great Chesterford (Essex); Hinxton, Duxford, Whittlesford, Sawston, Stapleford, Great Shelford, Trumpington, Cambridge, Fen Ditton, Chesterton, Milton, Waterbeach, Stretham, Thetford, Ely (Cambs.); Mildenhall, Lakenheath, Brandon (Suffolk); Weeting with Bromehill (Norfolk); Santon Downham (Suffolk); Thetford (Norfolk).
[Ely to Peterborough:] From Ely, through Witchford, Downham, Coveney, Witcham, Chatteris, Doddington, Benwick, Whittlesey (Cambs.); Stanground (Cambs. & Hunts.); Woodstone, Fletton (Hunts.); terminating at Fletton adjoining Peterborough (Northants.).
[Branch to Wisbech:] From Chatteris, through Doddington, Wimblington, March, Elm; terminating in Wisbech (Cambs.).
(b) Another plan, deposited 30 Nov. 1843. Map 1824.
There is no copy of this plan in the Northants. series.

Direct Northern Railway from London to York by way of Lincoln.
(a) Book of reference, deposited 30 Nov. 1844. Parts I–V. ML 666.
(b) Another copy of Parts I–II, no cover. ML 675.
(c) Another copy of Parts III–IV, no cover. ML 674.
(d) Another copy of Part V, no cover. ML 668.
(e) Plan, with memorial. Map 1801.
(f) Plan, with memorial. Map 1813.
(g) Plan, no memorial. Map 1799.
(h) Plan, no memorial. Map 1800.
(i) Plan, no memorial. Map 1869.
(j) Sections, Wakefield Branch, lacking cover. Map 1848.

(k) Cover of plan of Part 1, London to Grantham, deposited 30 Nov. 1844, with some sheets of the plan, and other fragments, some of which appear to be from the London & York scheme of 1844, others not identified. Map 1903.
Northants. Main Series 13.

Midland Railways Extensions. Syston & Peterborough Railway.
(a) Book of reference, deposited 30 Nov. 1844. ML 670.
(b) Plan, with memorial. Map 1850.
(c) Plan, no memorial. Map 1849.
Northants. Main Series 15.

Cambridge & Lincoln Railway.
(a) Book of reference, lacking cover or memorial. ML 676.
(b) Section, deposited 30 Nov. 1844. Map 1882.
(c) Section, with memorial. Map 1911.
(d) Plan, with memorial. Map 1910.
(e) Plan, with memorial. Map 1904.
Northants. Main Series 16.

Boston, Stamford and Birmingham Railway, from Birmingham to Boston and Wisbech, with a branch to Market Harborough.
(a) Plan, deposited 30 Nov. 1845. Map 1902.
(b) Plan, with memorial. Map 1851.
(c) Section, with memorial. Map 1852.
Northants. Main Series 19.

Direct Northern Railway from London to York, with branches.
(a) Plan, deposited 29 Nov. 1845. Map 1812.
(b) Plan, no memorial. Map 1798.
(c) Book of reference. ML 677.
Northants. Main Series 24.

London & York (Great Northern) Railway. Stamford & Spalding Branch.
(a) Plan, deposited 30 Nov. 1845. Map 1845.
(b) Plan, no memorial. Map 1845.
Northants. Main Series 27.

Stamford, Market Harborough & Rugby Railway, with branch to the Leicester & Bedford Railway.
(a) Book of reference, deposited 30 Nov. 1845. ML 667.
(b) Another copy, lacking cover and first four pages. ML 673.
(c) Plan, with memorial. Map 1897.
(d) Sheets 1–4, 8–9 of plan, lacking covers. Map 1889.
Northants. Main Series 28.

Peterborough, Spalding & Boston Junction Railway. 1845.
(a) Plan, deposited 30 Nov. 1845.
(b) Cover (only) of plan, no memorial. Map 1884.
(c) Sheets 2–6 and [7] of plan, lacking cover. Map 1889.
(d) 1:63,360 Ordnance Survey map showing line of proposed railway in red, terminating at the turnpike road west of Spalding, where it makes an end-on junction with a railway continuing towards Boston, marked in black ink. Map 1804.
Northants. Main Series 35.

Peterborough & Nottingham Junction Railway.
(a) Plan, deposited 30 Nov. 1845. Map. 1886.
(b) Plan, with memorial. Map 1818.
Northants. Main Series 40.

Midland Railway. Branches and deviations from the Syston & Peterborough Railway.
(a) Plan, deposited 30 Nov. 1845. Map 1884.
(b) Large number of disbound sheets, no cover, some (possibly all) from the plan for this line. Map 1848.
Northants. Main Series 41.

Port of Wisbech, Peterborough, Birmingham & Midland Counties Union Railway.
(a) Plan, deposited 30 Nov. 1845.
Northants. Main Series 44.

Lynn, Wisbeach & Peterborough, Midland Counties' and Birmingham Junction Railway.
(a) Plan, deposited 30 Nov. 1845. Map 1884.
(b) Plan, with memorial. Map 1880.
(c) Plan, no memorial. Map 1810.
(d) Sections, lacking cover. Map 1848.
Northants. Main Series 45.

Peterborough, Wisbeach & Lynn Junction Railway.
(a) Book of reference, deposited 30 Nov. 1845. Map 1854.
(b) Plan, no memorial. Map 1854.
Northants. Main Series 46.

Eastern Counties Railway. York Extension. Line between Cambridge and Lincoln, with branches to Sleaford, Boston and Spalding.
(a) Book of reference, no memorial. ML 672.
(b) Plan, deposited 29 Nov. 1845. Map 1797.
(c) Plan, no memorial. Map 1796.
Northants. Main Series 49.

Great Northern Railway. Deviations between London and Grantham.
(a) Plan, no memorial (Northants. counterpart deposited 28 Nov. 1846). Map 1791.
(b) Plan, no memorial. Map 1870.
(c) 1:63,360 Ordnance Survey map, on which has been marked the parliamentary line of the Great Northern Railway and deviations between London and Grantham. Map 1830.
Northants. Main Series 52.

Boston, Stamford & Birmingham Railway. Peterborough & Thorney Line.
(a) Plan, no memorial. Map 1794.
(b) Plan, no memorial. Map 1872.
(c) 1:63,360 Ordnance Survey map showing line from Peterborough to Thorney; also the Boston, Stamford & Birmingham Railway (Stamford & Wisbech Line); a Harbour Branch at Wisbech; and a line from Wisbech to Sutton. Map 1839.
(d) Duplicate of (c). Map 1809.
Northants. Main Series 54.

Drainage of Crowland and Cowbit Washes and other lands. 1846.
(a) Plan, no memorial. Map 1887.
(b) Plan, no memorial. Map 1900.
Northants Main Series. 57.

[Midland Railway.] Syston & Peterborough Railway. Deviations.
(a) Plan, no memorial; Northants. counterpart deposited 30 Nov. 1846. Map 1795.
(b) Plan, no memorial. Map 1888.

(c) Cover (only) from another plan. Map 1889.
Northants. Main Series 58.

Great Northern Railway. Deviations between Peterborough, Doncaster and Boston. 1846.
(a) Book of reference. ML 669.
(b) Plan, no memorial. Map 1847.
(c) Plan, no memorial. Map 1828.
(d) 1:63,360 Ordnance Survey map, on which has been marked the parliamentary line of the Great Northern Railway; deviations between Peterborough, Doncster and Boston; and other proposed deviations. Map 1806.
Northants. Main Series 59.

River Nene. Improvement and Drainage. 1847.
(a) Plan, deposited 30 Nov. 1847. Map 1866.
Northants. Main Series 62.

Nene Valley Drainage and Navigation Improvement. 1851.
(a) Plan, no memorial.
Northants. Main Series 64.

Nene Valley Drainage & Improvements.
(a) Book of reference, deposited 30 Nov. 1858. Map 1875.
(b) Plan, cover torn, probably with loss of memorial. Map 1898.
(c) Plan, deposited 30 Nov. 1858. Map 1901.
Northants. Main Series 72.

Stamford & Essendine Railway.
(a) Book of reference, no memorial, with *Gazette* notice dated 12 Nov. 1861.
(b) Plan, deposited 30 Nov. [1861].
Northants. Main Series 76.

Eastern Counties Railway. Railways between Wisbech and Peterborough.
(a) Plan, deposited 30 Nov. 1861. Map 1867.
(b) Plan, no memorial. Map 1867.
Northants. Main Series 79.

Nene Valley Drainage and Navigation Improvements. John Fowler, engineer.
(a) Plan, deposited 30 Nov. 1861. Map 1829.
(b) Plan, no memorial. Map 1899.
(c) Book of reference, no memorial. Map 1877.
Northampton Main Series 115.

Peterborough, Wisbech & Sutton Railway.
(a) Plan, deposited 29 Nov. 1862. Map 1825.
Northants. Main Series 81.

Great Eastern Northern Junction Railway.
(a) Plan, deposited 30 Nov. 1863. Map 1905.
(b) Plan, with memorial. Map 1885.
Northants. Main Series 89.

Stamford & Essendine Railway. Sibson Extension.
(a) Plan, deposited 25 Nov. 1863.
(b) Plan, with memorial. Map 1856.
Northants. Main Series 90.

Lanashire & Yorkshire and Great Eastern Junction Railway.
(a) Plan, deposited 30 Nov. 1864. Map 1792.
(b) Plan, no memorial. Map 1814.
Northants. Main Series 97.

Peterborough Waterworks.
(a) Plan, deposited 30 Nov. 1865. Map. 1846.
(b) Plan, with memorial. Map 1892.
Northants. Main Series 109.

Peterborough Water.
(a) Plan, deposited 30 Nov. 1866. Map 1860.
Northants. Main Series 111.

Peterborough Gas.
(a) Two plans, one with memorial, one without, deposited 29 Nov. 1867. Map 1859.
Northants. Main Series 113.

Coal Owners' Associated London Railway.
(a) Plan and book of reference, deposited 30 Nov. 1870. Map 1811.
(b) Plan, with memorial. 1870. Map 1908.
Northants. Main Series 127.

Great Northern Railway. Additional Powers.
(a) Plan, deposited 29 Nov. 1872. Map 1842.
(b) Plan, deposited 29 Nov. 1872. Map 1841.
Northants. Main Series 139.

Peterborough Water.
(a) Plan, deposited 30 Nov. 1874. Map 1816.
(b) Plan, no memorial. Map 1896.
(c) Two books of reference, neither with memorial, kept with Maps 1816 and 1896.
Northants. Main Series 150.

Braceborough Water. 1876.
(a) Plan, deposited 30 Nov. 1875. Map 1855.
(b) Plan and book of reference, with memorial. Map 1857.
Northants. Main Series 156.

Great Northern Railway.
(a) Two sets of plan and book of reference, deposited 29 Nov. 1876. Map 1817.
Northants. Main Series 158.

Great Eastern Railway (Northern Extension).
(a) Book of reference. ML 671
(b) Plan, deposited 30 Nov. 1877. Map 1793.
(c) Plan, deposited 30 Nov. 1877. Map 1881.
Northants. Main Series 160.

Market Deeping Railway.
(a) Plan, deposited 29 Nov. 1877. Map 1858.
Northants. Main Series 162.

Peterborough Tramways.
(a) Two plans, one with memorial, one without, deposited 29 Nov. 1879. Map 1907.
Northants. Main Series 172.

INDEX

This index includes all the place-names, the names of companies and other undertakings, and the names of engineers and surveyors listed in the Northamptonshire sections of the catalogue. The summary list of Peterborough plans has not been indexed. All numbers refer to plans, not pages; canal plan numbers are prefixed with C, turnpike road plans with T.

Ab Kettleby (Leics.) 40, 86, 91, 132, 135
Abbots Langley (Herts.) 1, 3, C7
Abbots Morton (Worcs.) 22, 85
Abbots Ripton (Hunts.) 13
Abbots Salford (Warwicks.) 85
Abbotsley (Hunts.) 29, 39
Abington (Northants.) 25, 29, 31, 39, 64, 104, 122, T5
    electricity 200
Abthorpe (Northants.) 21, 34, 47, 55, 84, 120, 151, 189, C24
    electricity 294
Acaster Malbis (Yorks.) 6, 13, 24
Acaster Selby (Yorks.) 6, 13
Acton (Middlesex) 3, 262
Adderbury (Oxon.) 7, 11, 14, 92, 94a, 101, 102, 105, 125, 138, 140, 261, 313
Addington (Bucks.) 30, 51
    see also Great Addington; Little Addington
Addison, John 36, 88
Addy, John 172
Adstock (Bucks.) 30, 51
Adstone (Northants.) 85, 94
    electricity 294
Adwell (Oxon.) 30
Adwick le Street (Yorks.) 13, 24
Ailsworth (Northants.) 10, 64, 75, 90
Aisthorpe (Lincs.) 97, 160
Akeley (Bucks.) 51, T6
Albano, B. 32, 44
Albury (Oxon.) 32
Alcester (Warwicks.) 22, 85
Alconbury (Hunts.) 13, 24, 33, 42, 56
Aldbury (Herts.) 3, 38, C7
Alderminster (Worcs.) 14, 22, 85, 94
Alderton (Northants.) 30, 51
Aldington (Worcs.) 14
Aldwincle (Northants.) 64, T14
    electricity 291
Alexander & Littlewood 86, 95, 102
Alexander & Ross 232, 236
    see also Ross, Alexander
Algarkirk (Lincs.) 13, 19, 49
Allesley (Warwicks.) 1, T8B
Allexton (Leics.) 86, 88, 91, 137, 142
Alne: see Great Alne
Alsager (Cheshire) 23
Althorp (Northants.) 149
    electricity 290
Alvechurch (Worcs.) 36
Alveston (Warwicks.) 22, 29
Alwalton (Hunts.) 10, 64
Amblecote (Staffs.) 14
Ambrosden (Oxon.) 14, 32
Ampthill (Beds.) 23
Ancaster (Lincs.) 13

Annesley (Notts.) 26, 209
Anstey Pastures (Leics.) 209
Anston, North & South (Yorks.) 135
Ansty (Warwicks.) C25
Anwick (Lincs.) 6, 16, 49, 97
Apethorpe (Northants.) 64, 141
Appleton Roebuck (Yorks.) 6
Appletree (Northants.) 94
Archer: see Sherman & Archer
Ardley (Oxon.) 14, 32, 262
Arksey: see Bentley with Arksey
Arlesey (Beds.) 13, 24, 42
Armston (Northants.) 64
    electricity 291
Armthorpe (Yorks.) 89, 97
Arncot (Oxon.) 32
Arrow (Warwicks.) 22, 85
Arthingworth (Northants.) 8, 33, 42, 56, 66, 141, 144, 245, 246
    electricity 296
Asfordby (Leics.) 15, 132, 137
Asgarby (Lincs.) 6, 16, 49, 86, 91
Ashby: see also Castle Ashby; Cold Ashby
Ashby de la Launde (Leics.) 89, 97, 127, 137
Ashby de la Zouch Canal C1
Ashby de la Zouch (Leics.) 23, C1
Ashby Folville (Leics.) 91, 86
Ashby Magna (Leics.) 209, 213
Ashby St Ledgers (Northants.) 1, 3, 4, 5, T2
    electricity 294
Ashby Woulds (Leics.) 23
Ashchurch (Gloucs.) 47
Ashendon (Bucks.) 30, 262
Ashley (Northants.) 8, 19, 28, 33, 37, 86, 88, 91, 135, 137, 142, C13, C14
    electricity 296
Ashton [by Oundle] (Northants.) 10, 64
    electricity 291
Ashton [by Roade] (Northants.) 1, 3, 149
    gas 309
Ashton Underhill (Worcs.) 47
Ashwell (Rutland) 15, 41, 58
Askern (Yorks.) 97
Askham (Notts.) 13
Askham Bryan (Yorks.) 6
Aslackby (Lincs.) 6, 16, 49, 60, 89, 97, 127, C15
Aston (Warwicks.) 1, 3, 48
    see also Cold Aston; North Aston
Aston Bampton (Oxon.) 313
Aston Cantlow (Warwicks.) 36
Aston Clinton (Bucks.) 38
Aston Flamville (Leics.) T2
Aston le Walls (Northants.) 22, 85, 94
Aston Somerville (Worcs.) 47
Aston Subedge (Gloucs.) 47

Astwell with Falcutt (Northants.) 47, 55, 84, 95, 120, C24
Aswarby (Lincs.) 97
Atherstone on Stour (Warwicks.) 14, 85, 94
Attercliffe cum Darnall (Yorks.) 13
Attington (Oxon.) 30
Audley (Staffs.) 23
Ault Hucknall (Derbys.) 209
Aunby (Lincs.) 52
Austerfield (Yorks.) 13, 59
Aylesbury (Bucks.) 14, 18, 38, 51, 53, 196, 203
Aylestone (Leics.) 8, 209, 213, C4
Aynho (Northants.) 7, 11, 14, 32, 262, 270, T10, T12
Ayston (Rutland) 132

Babworth (Notts.) 13, 135
Badby (Northants.) 29, 36, T13, T15
    electricity 290
Badsey (Worcs.) 14
Badsworth (Yorks.) 24
Bainton (Northants.) 13, 27, 41, 52, 75, 90, 215, 219, 225, 268
Baker, William 48, 77, 82, 118, 137, 141, 142, 144, 147, 149, 153, 159, 163, 167
    see also Barlow, Son & Baker
Bakewell: see Cooper, R. Elliott, Leane & Bakewell; Leane & Bakewell
Balby with Hexthorpe (Yorks.) 59, 89, 135
Balderton (Notts.) 13, 24
Balne (Yorks.) 6, 13
Bampton (Oxon.) 313
Banbury (Oxon.) 14, 21, 34, 38, 47, 68, 92, 95, 96, 98, 102, 125, 140, 226
    gas 308
    rural district 313
    water 96
    see also Warkworth
Banbury & Cheltenham Direct Railway 138
Banbury Gaslight & Coke Co. 308
Barby (Northants.) 3, 4, 5, 14, 149, 209, 213, 218, C12, C21, C25
Barcheston (Warwicks.) 92
Bardney (Lincs.) 13
Barford (Northants.) 42, 56, 67, 132, 135, 136, 165
    gas 304
    see also Little Barford
Barford St John (Oxon.) 94a
Barholm (Lincs.) 13, 49, C15
Barkston (Lincs.) 13, 24, 31
Barleythorpe (Rutland) 43
Barlings (Lincs.) 31
Barlow, Son & Baker 154, 165
Barlow, William Henry 135, 145
Barnack (Northants.) 13, 15, 19, 24, 41, 52,

58, 75, 90, 109, 178
electricity 297
Barnby Dun (Yorks.) 13, 89, 97, 160
Barnby Moor (Notts.) 13, 135
Barnby in the Willows (Notts.) 24
Barnet: *see* East Barnet; Friern Barnet
Barnston: *see* Langar cum Barnston
Barnwell All Saints (Northants.) 10, 64
electricity 291
Barnwell St Andrews (Northants.) 10, 64
electricity 291
Barrow (Rutland) 31, 40
Barrow upon Soar (Leics.) 209, 213
Barrowden (Rutland) 28, 33, 141
Barston (Warwicks.) 3
Bartholomew, Charles 127, 155
Barthomley (Cheshire) 23
Barton (Cambs.) 29, 39, 127
Barton in the Clay (Beds.) 23
Barton Hartshorn (Bucks.) 18, 196, 203, 209, 213
Barton Seagrave (Northants.) 20, 23, 33, 43, 56, 67, 80, 136, 247
  Burton Latimer water 273
  electricity 289
  gas 304
Bascote (Warwicks.) C24
Basford (Notts.) 26, 209, 213
Basford (Staffs.) 23
Basing (Hants.) 30
Basingstoke (Hants.) 30
Bassingthorpe (Lincs.) 13, 24, 52
Baston (Lincs.) 6, 16, 49, 60, 89, 97, 127, 150, 156, C15
Battlesden (Beds.) T8B
Bawtry (Yorks.) 13
Bazalgette, J.W. 16
Beachampton (Bucks.) C5
Beaconsfield (Bucks.) 18, 32
Beanfield Lawns (Northants.), electricity 296
  gas 304
Bearley (Warwicks.) 36
Bearn, G.F. 257
Beaumont Leys (Leics.) 209
Beausale (Warwicks.) C24
Beckford (Worcs.) 47
Beckingham (Lincs.) 13, 24
Beckingham (Notts.) 6, 13, 24, 59, 89
Bedford (Beds.) 13, 20, 23, 24, 25, 42, 43, 56, 67, 103, 104, 136
Bedford & Northampton Railway 106, 110
Bedford, Northampton & Leamington Railway 103
Bedford, Northampton & Weedon Railway 99
Beech Hill (Berks.) 30
Beer, Walter 231
Begbrooke (Oxon.) 11
Belgrave (Leics.) 213
Belton (Lincs.) 13, 24, 31
Benefield (Northants.), electricity 291
Bengeworth (Worcs.) 14
Bentley with Arksey (Yorks.) 13, 24, 59, 135
Berkhampstead (Herts.) 3, C7
Berkswell (Warwicks.) 1, 3, C21
Bernard, Charles E. 30
Bevan, B. C10, C12, C13, C14
Bicester (Oxon.) 14, 32, 313, 14, 262
  urban district 313
Bickenhill (Warwicks.) 1, 3, T8B
Biddenham (Beds.) 20, 23, 25, 42, 56, 67, 103, 104, 136
Bidder, George P. 66
Biddlesden (Bucks.) 189, 199, 234
Bidford on Avon (Warwicks.) 22, 85, 94
Bierton with Broughton (Bucks.) 14, 38, 51, 53
Biggin: *see* Newton & Biggin
Biggleswade (Beds.) 13, 24, 275

Billesdon (Leics.) 137, 142
Billing: *see* Great Billing; Little Billing
Billingborough (Lincs.) 6, 16, 49, 89, 97, 127
Billinghay (Lincs.) 13, 16
Bilsthorpe (Notts.) 135
Bilton (Warwicks.) 1, 3, 11, C21
Bingham (Notts.) 135, 137, 142
Binley (Warwicks.) 1, 3
Binnie, Deacon & Gourley 318
Binns, J.G. 56
Binton (Warwicks.) 22, 85, 94
Birch, E. 34
  J.B. 34
Birdingbury (Warwicks.) 7, 50, 202, C6
Birkin (Yorks.) 6, 13
Birmingham (Warwicks.) 1, 3, 48, 300, T8B
Birmingham & Gloucester Railway 36
Birstall (Leics.) 213
Birthorpe (Lincs.) 97, 127
Bisbrooke (Rutland) 31, 43, 132, 145
Bishops Itchington (Warwicks.) 11, 29, 50
Bishopthorpe (Yorks.) 13, 24
Bitchfield (Lincs.) 13
Bitteswell (Leics.) T2
Bix (Oxon.) 30
Blackfordby (Leics.) 23
Blackthorn (Oxon.) 14, 262
Blackwell (Derbys.) 26, 209
Blakesley (Northants.) 85, 94
  electricity 294
Blankney (Lincs.) 6, 13, 16, 49, 97, 127, 160
Blaston (Leics.) 86, 88, 91, 135, 137, 142
Blatherwycke (Northants.) 141
Blaxton (Yorks.) 89
Bledington (Gloucs.) 98, 138
Bledlow (Bucks.) 32
Bletchington (Oxon.) 7, 11
Bletchley (Bucks.) 1, 3, 61, 149
  electricity 290
Bletsoe (Beds.) 29, 39, 43
  electricity 287
Blisworth (Northants.) 1, 3, 10, 30, 34, 47, 51, 55, 84, 85, 95, 104, 129, 134, 148, 152, 155, 189, 317, C7
  electricity 290
Blockley (Gloucs.) 47, 68, 92, 102, 125, 140
Blockley & Banbury Railway 92, 140
Bloxham (Oxon.) 34, 47, 68, 92, 94a 98, 101, 102, 105, 125, 138, 140
Bloxholm (Lincs.) 127
Blunham (Beds.) 13
Bluntisham cum Earith (Hunts.) 6, 89, 97, 127, 160
Blyth (Notts.) 13, 59, 135
Blyton (Lincs.) 97, 160
Boarstall (Bucks.) 32
Bobbington (Staffs.) 262
Boddington, Upper & Lower (Northants.) 22, 85, 94, C24
Bodicote (Oxon.) 14, 92, 94a 98, 101, 102, 105, 125, 140
Bole (Notts.) 6, 24, 89
Bolnhurst (Beds.) 29, 39
Bolton Percy (Yorks.) 6
Boothby Graffoe (Lincs.) 13, 31, 89
Boothby Pagnell (Lincs.) 13
Borough Fen (Northants.) 35, 59, 97, 100, 160, 210, C14
Borthwick, M.A. 60
Bosley (Cheshire) 23
Boston (Lincs.) 13, 19, 49, 59, C15
Boston, Stamford & Birmingham Railway 19, 54
Bothamsall (Notts.) 135
Boughton (Northants.) 8, 42, 56, 64, 66, 157, 186
  electricity 258
  Northampton water 281, 282, 316

Boultham (Lincs.) 6, 13, 24, 89, 160
Bourn (Cambs.) 29, 39
Bourne (Lincs.) 6, 16, 49, 60, 89, 97, 127, 279, C15
Bourton (Oxon.) 7, 11, 14
Bourton, Chipping Norton & Banbury Railway 98
Bourton on the Water (Gloucs.) 138
Bovill, George 109
Bovingdon (Herts.) C7
Bow: *see* Stratford le Bow
Bowden: *see* Great Bowden; Little Bowden
Bozeat (Northants.) 29, 39, 61, 108, 119
  electricity 288
  gas 312
  tramroads 198, 205
Braceborough (Lincs.) 13, 150, 156, 219
  water 156
Bracebridge (Lincs.) 31, 89
Brackley (Northants.) 38, 53, 151, 196, 203, 209, 213, 218, 221, 313
  electricity 293, 294
  rural district 313
  electricity 294
Bradden (Northants.) 21, 47, 55, 84, 85, 94, 120, 151, C24
  electricity 294
Bradenham (Bucks.) 32, 51
Bradwell (Bucks.) 1, 3, 149, C7, T4
Bradwell Abbey (Bucks.) 1, 3, 149, T8B
Brafield on the Green (Northants.) 10, 25, 29, 39, 64, 104, T7
  electricity 290
  gas 292
Brailes (Warwicks.) 47, 68, 92, 102, 125, 140
Braithwell (Yorks.) 13, 135
Bramley (Hants.) 30
Bramley (Yorks.) 13
Brampton (Hunts.) 13, 24, 42, 56, 87
Brampton (Lincs.) 6
  see also Chapel Brampton; Church Brampton
Brampton Abbotts (Herefs.) 125
Brampton Ash (Northants.), electricity 289
Brampton en le Morthen (Yorks.) 13
Bramwith: *see* Kirk Bramwith
Brandon & Bretford (Warwicks.) 7
Branston (Leics.) C4
Branston (Lincs.) 6, 13, 16, 49, 127, 160
Brattleby (Lincs.) 97, 160
Brauncewell (Lincs.) 89, 127
Braunston (Northants.) 50, 74, 93, 103, 190, 202, 209, 213, 216, 300, C6, C7, C8, C12, C21, C25, T1, T8B
  electricity 294
Braunston (Rutland) C15
Braunston Frith (Leics.) 23
Braunstone (Leics.) 23
Braybrooke (Northants.) 20, 23, 28, 33, 42, 56, 66, 67, 284
  electricity 289
Brayfield: *see* Cold Brayfield
Brayton (Yorks.) 6, 13, 24
Bredicot (Worcs.) 85
Brentford: *see* New Brentford
Brentingby & Wyfordby (Leics.) 15, 40, 41, 58, 132, 137
Bretford: *see* Brandon & Bretford
Bretforton (Worcs.) 14, 29
Briant, Nathaniel 43
Brickhill: *see* Great Brickhill; Little Brickhill
Bridgford: *see* East Bridgford; West Bridgford
Bridgnorth (Salop) 262
Briggs, James 284
Brightside Bierlow (Yorks.) 13
Brigstock (Northants.)
  electricity 291, T14

INDEX

gas 304
Brill (Bucks.) 262
Bringhurst (Leics.) 8, 19, 28, 33, 37, 132, 137, 142
Brington (Hunts.) 33, 56
Brington (Northants.) 149, 167, 217
  electricity 290
Briningham (Norfolk) 276
Brinklow (Warwicks.) C21, C25
Brinsworth (Yorks.) 13
British Electric Traction Co. Ltd 239
British Electric Traction Ltd 253, 260
British Transport Commission 317
Brixworth (Northants.) 8, 42, 56, 66
  electricity 290
  Northampton water 316
Broadholme (Notts.) 6
Broadway (Worcs.) 47
Brockhall (Northants.) 3, 4, 78, 112, 185, C7, T2
  electricity 290
  gas 305
Bromham (Beds.) 20, 23, 42, 56, 67, 99, 103, 104, 136
Bromsgrove (Worcs.) 36
Brooke (Rutland) C15
Brooksby (Leics.) 15, 41
Brothertoft (Lincs.) 13
Broughton (Bucks.) 14
Broughton (Hunts.) 16, 49
Broughton (Northants.) 20, 23, 31, 33, 67, T5
  electricity 289
  gas 304
  Northampton water 283
Broughton (Oxon.) 34, 47, 68, 92, 102, 125, 140
  see also Bierton with Broughton; Nether Broughton; Upper Broughton
Broughton Hackett (Worcs.) 85
Brown, Charles J. 278, 279, 301
  Leonard H. 316
Brownsover (Warwicks.) 218, C25
  Rugby water 285
Broxholme (Lincs.) 13, 89, 97, 160
Bruce, George B. 80, 81, 100, 135
  George C. 87
Brunel, I.K. 11
Brunlees, James 80, 87
Brydone, Walter Marr 76, 90
Buck, G.W. 16
Buckby: see Long Buckby
Buckden (Hunts.) 13, 24, 87
Buckingham (Bucks.) 30, 38, 51, 151, 189, 199, 234, 313, C5, T3, T4, T6
  electricity 294
  rural district 313
Buckingham & Brackley and Oxford & Bletchley Junction Railway 53
Buckingham, Towcester & Metropolitan Junction Railway 234
Buckinghamshire & Northamptonshire Railways Union Railway 151
Buckinghamshire Railway 38, 120
Buckland (Bucks.) 38
Buckland (Gloucs.) 47
Bucknall (Lincs.) 13
Bucknell (Oxon.) 14, 32, 262
Budbrooke (Warwicks.) 36, 300, C6
Budby: see Perlethorpe cum Budby
Bugbrooke (Northants.) 1, 3, 29, 39, 48, 99, 103, 124, 133, 185, 190, C7, T8A, T8B
  electricity 290
  gas 292
Bulkeley (Cheshire) 23
Bulwell (Notts.) 26, 209, 213
Burghfield (Berks.) 30
Burghwallis (Yorks.) 13, 24, 89, 97, 135, 160

Burke, James B. 94, 101, 104, 105, 107, 128, 134
  Joseph F. 128, 148
Burley (Rutland) 15, 31, 41
Burmington (Warwicks.) 47, 68, 92, 102, 125, 140
Burn (Yorks.) 13
Burnage (Lancs.) 23
Burnham (Bucks.) 32, 262
Burrough on the Hill (Leics.) 86, 91, 135, 137
Burslem (Staffs.) 23
Burton: see also Gate Burton; West Burton
Burton by Lincoln (Lincs.) 13, 89, 97, 160
Burton Coggles (Lincs.) 13, 24, 52
Burton Dassett (Warwicks.) 14, 22, 50, 85, 94
Burton Hastings (Warwicks.) T2
Burton Latimer (Northants.) 23, 33, 42, 43, 56, 67, 80, 136, 175
  electricity 288, 289
  gas 304
  water 273
Burton Latimer Gas Co. Ltd 314
Burton Lazars (Leics.) 43, 86, 91, 137
Burton Overy (Leics.) 8, 19, 20, 23, 42, 56, 67, C4
Burton Pedwardine (Lincs.) 6, 89, 97, 127, 160
Burton upon Trent (Staffs.) 23
Bury (Hunts.) 16, 49, 89, 97, 127, 160
Bushbury (Staffs.) 262
Bushey (Herts.) 3
Bushley (Worcs.) 125
Buslingthorpe (Lincs.) 31
Butlers Marston (Warwicks.) 14, 22, 85, 94
Butley (Cheshire) 23
Byfield (Northants.) 22, 85, 94, 134, 313, C24
  electricity 294
Byrne, Oliver 51
Bytham: see Castle Bytham; Little Bytham
Bythorn (Hunts.) 33, 42, 56

Caddington (Beds.) T8B
Caldecote (Cambs.) 29, 39
Caldecote (Hunts.) 13, 24
Caldecote (Warwicks.) 14, 36, 93, 103
Caldecott (Rutland) 8, 19, 28, 31, 33, 37, 132
  Corby water 299
  see also Chelveston cum Caldecott
Calverton (Bucks.) C5, T8B
Camblesforth (Yorks.) 13
Cambridge (Cambs.) 6, 16, 29, 39, 49, 127
Cambridge & Lincoln Railway 16
Cammeringham (Lincs.) 160
Campsall (Yorks.) 6, 13, 24, 97, 160
Campton (Beds.) 20, 67
Canons Ashby (Northants.) 85, 94, 101, 105, 196, 203, 209, 213, 226
  electricity 294
Cantley (Yorks.) 6, 13, 24, 59, 89, 135
Canwick (Lincs.) 6, 13, 49, 89, 160
Cardington (Beds.) 13, 24, 42, 67
Careby (Lincs.) 13, 24, 52, 275
Carlby (Lincs.) 13, 24, 52
Carleton (Yorks.) 13
Carlton & Chellington (Beds.), gas 309
  see also East Carlton; North Carlton; South Carlton
Carlton on Trent (Notts.) 13
Carlton Scroop (Lincs.) 13
Carr House & Elmfield (Yorks.) 89, 135
Castle Ashby (Northants.) 10, 29, 39, 64, 104, T7
  electricity 290
  gas 307
Castle Bytham (Lincs.) 24, 52
Castle Gresley (Derbys.) 23

Castleford (Yorks.) 24
Castlethorpe (Bucks.) 1, 3, 149
Castor (Northants.) 10, 13, 24, 58, 64, 75, 90, 109
Catesby (Northants.) 29, 209, 213, 218, 233
  electricity 294
Catthorpe (Leics.) 33, 37, 167
  Rugby water 285
Catworth (Hunts.) 33, 42, 56, 87
Caversham (Oxon.) 30, 51
Caverswall (Staffs.) 23
Cawood (Yorks.) 6, 13, 24
Caxton (Cambs.) 29, 39
Caythorpe (Lincs.) 13, 31
Caythorpe (Notts.) 135
Chaceley (Gloucs.) 125
Chackmore: see Radclive cum Chackmore
Chaddesley Corbett (Worcs.) 36
Chadshunt (Warwicks.) 85, 94
Chalcombe (Northants.) 7, 14, 34, 85, 95, 101, 105, 226, 313
Chalfont St Giles (Bucks.) 18
Chalfont St Peter (Bucks.) 18
Chalgrove (Beds.) T8B
Chaloner, Mr T3
Chalvey: see Upton cum Chalvey
Chapel Ascote (Warwicks.) 11, 29
Chapel Brampton (Northants.) 64, 66, 186, C4
  electricity 258
  Northampton water 281, 283
Chapel Haddlesey (Yorks.) 6
Charlbury (Oxon.) 313
Charlecote (Warwicks.) 29
Charlton Kings (Gloucs.) 138
Charwelton (Northants.) 22, 134, 209, 213, 223, 233, 311
  electricity 294
Cheadle (Cheshire) 23
Cheadle (Staffs.) 23
Checkley (Staffs.) 23
Cheddington (Bucks.) 3, 14, C7
Cheddleton (Staffs.) 23
Cheffins, C.F. & Son 79
  Charles F. 10, 53, 60, 61, 68
  Charles R. 92, 97, 99, 100, 125, 129, 140
Chellington: see Carlton & Chellington
Chelsea (Middlesex) 1, 3
Cheltenham (Gloucs.) 138
Chelveston cum Caldecott (Northants.) 10, 64, 201, 211
  electricity 277, 287
  gas 307
  tramroads 198, 239, 253, 260
Chepping Wycombe (Bucks.) 32, 51
Cherington (Warwicks.) 47, 68, 92, 102, 125, 140
Cherry Willingham (Lincs.) 13, 31, 97, 127
Chesterfield (Derbys.) 26
Chesterton (Cambs.) 6, 16, 49, 127
Chesterton (Hunts.) 13, 24, 64
Chesterton (Oxon.) 14
Chesterton & Kingston (Warwicks.) 29, 103
Chetwode (Bucks.) 18, 196, 203, 209, 213
Chicksands Priory (Beds.) 20
Chilton (Bucks.) 32
Chilworth (Oxon.) 32
Chinnor (Oxon.) 32
Chipping Barnet (Herts.) 23
Chipping Campden (Gloucs.) 47, 125
Chipping Norton (Oxon.) 34, 94a, 98, 101, 102, 105, 313
  rural district 313
Chipping Norton & Banbury Railway 94a
Chipping Norton, Banbury & East & West Junction Railway 101
Chipping Norton, Banbury Railway & East & West Junction Railway 105
Chipping Warden (Northants.) C24
Chorlton upon Medlock (Lancs.) 23

INDEX

Church Brampton (Northants.) 42, 56, 64, 66, 149
   electricity 258
Church Gresley (Derbys.) 23
Church Langton (Leics.) 19, 37, 42, 56, 67
Church Lawford (Warwicks.) 1, 3, C25
Church Lawton (Cheshire) 23
Churchill (Oxon.) 34, 98, 138
Churchill (Worcs.) 85
Churchover (Warwicks.) 218, T2
Claines: *see* North Claines
Clapham (Beds.) 20, 23, 43, 67
Clarborough (Notts.) 24
Clark, John 277
Clarke, William 118
Clattercote (Oxon.) 7, 11, 14, C24
Claverdon (Warwicks.) 36
Claverley (Salop) 262
Clawson: *see* Long Clawson
Claybrook (Leics.) T2
Claydon (Oxon.) 7, 11, 14, 85, 94, C24
   see also East Claydon; Middle Claydon; Steeple Claydon
Claypole (Lincs.) 13, 24
Cleeve Prior (Worcs.) 85
Clenchwarton (Norfolk) 45
Clifford Chambers (Gloucs.) 14, 22, 29, 85, 94
Clifton (Beds.) 13, 24, 42
   see also North Clifton
Clifton Reynes (Bucks.) 25, 99, 103, 104, 110
   gas 309
Clifton on Dunsmore (Warwicks.) 3, 11, 33, 37, 149, 209, 213, 218, C25
   Rugby water 285
Clipston (Northants.) 28, 33, 66, C4
   electricity 296
Clopton (Northants.) electricity 291
Coal Owners' Associated London Railway 127
Coates (Cambs.) 64
Coates (Lincs.) 97
Cobley: *see* Tutnall & Cobley
Coddington (Notts.) 24
Coe & Mann 44
Cofton Hackett (Worcs.) 36
Cogenhoe (Northants.) 10, 29, 39, 64, 103
   electricity 290
   gas 292, 309
Cold Ashby (Northants.), C10
   electricity 294
Cold Aston (Gloucs.) 138
Cold Brayfield (Bucks.) 104, T7
   gas 309
Cold Higham (Northants.) T8A, T8B
   electricity 290
Cold Newton (Leics.) 86, 91, 137
Coleby (Lincs.) 13, 31, 89
Coleorton (Leics.) C1
Coleshill (Bucks.) 18
Collingtree (Northants.) 8
   electricity 258
Collister, John 84, 92, 95, 102
Collyweston (Northants.) 28, 33, C13, C14
   electricity 297
Colmworth (Beds.) 29, 39
Colne (Hunts.) 6, 89, 97, 127, 160
Colsterworth (Lincs.) 31
Colston Basset (Notts.) 43, 135
Comberton (Cambs.) 29, 39
Combrook (Warwicks.) 85, 94
Compton: *see also* Long Compton
Compton Abdale (Gloucs.) 138
Congingsby (Lincs.) 13
Conington (Hunts.) 13, 24
Conisbrough (Yorks.) 135
Consall (Staffs.) 23
Cooper, Cecil James 293
Cooper, R. Elliott, Leane & Bakewell 185

see also Elliott-Cooper, R.
Cople (Beds.) 13
Copmanthorpe (Yorks.) 6
Coppenhall: *see* Monks Coppenhall
Coppingford (Hunts.) 13, 24
Corby (Lincs.) 13, 24, 52
Corby (Northants.) 43, 145, 284
   electricity 296
   gas 304
   water 299
Corringham, Great & Little (Lincs.) 89, 97, 160
Corse (Gloucs.) 125
Cosby (Leics.) 209, 213
Cosford (Warwicks.) C25
Cosgrove (Northants.) 30, 107, 128, 192, C7, T8B
   gas 310, 315
Cotesbach (Leics.) 209, 213
Cotgrave (Notts.) 40, 43
Coton (Cambs.) 127
Coton (Northants.), electricity 290
   gas 305
   water 281
Cotterstock (Northants.) 10, 64
Cottesbrooke (Northants.) 42, 56, 66, C4
   electricity 290
Cottesmore (Rutland) 31, 40
Cottingham (Northants.) 19, 28, 31, 33, 37, 132, 137, 142, C13, C14
   electricity 296
   gas 304
Coughton (Warwicks.) 36, 85
Counthorpe (Lincs.) 52
Courteenhall (Northants.) 1, 3, 8, 149, 169, 216, 229
   electricity 290
Coventry (Warwicks.) 1, 3, 48, C21, T8B
Covington (Hunts.) 87
Cowbit (Lincs.) 57, 59, 100, 160
Cowick: *see* Snaith & Cowick
Cowley (Middlesex) 32, C7
Cowley (Oxon.) 32
Cranfield (Beds.) 107, 128
Cranford St Andrew (Northants.) 80
   electricity 296
   gas 304
Cranford St John (Northants.) 80
   electricity 296
   gas 304
Cranoe (Leics.) 137, 142
Cransley (Northants.) 20, 23, 31, 33, 157, 195, 220
   electricity 289
   gas 304
   Kettering water 243
Cranwell (Lincs.) 97, 127
Creaton: *see* Great Creaton
Creeton (Lincs.) 13, 24, 52
Crick (Northants.) 149, C10, C12, T2
   rural district, electricity 294
Cricklade (Wilts.) 313
Cricklade & Wootton Bassett (Wilts.), rural district 313
Crofton (Yorks.) 13
Cromwell (Notts.) 13
Cropredy (Oxon.) 7, 11, 14, 22, 94, 95, 226, C24
Cropthorne (Worcs.) 14, 85
Cropwell Bishop (Notts.) 40, 43, 135, 137, 142
Cropwell Butler (Notts.) 135, 137, 142
Crossley, John Sydney 83, 116, 123, 132, 136, 145
Croughton (Northants.) 18
Crowland (Lincs.) 13, 19, 27, 35, 57, 59, 100, 160, 210, C14
   drainage 57
Crowland Railway 210
Crowle (Worcs.) 22

Croxton (Cambs.) 29, 39
   see also South Croxton
Croxton Keyrial (Leics.) 132
Cubbington (Warwicks.) C24
Cubitt, Joseph 52, 59, 91
   William 52, 59
Cuddesdon (Oxon.) 32
Culworth (Northants.) 34, 85, 95, 101, 105, 226, 233, C24
Cutteslowe (Oxon.) 14, 53

Dalby: *see* Great Dalby; Old Dalby
Dallington (Northants.) 31, 42, 56, 64, 66, 129, 134, 149, 152, 163, 171, 175, 185, C4
   electricity 200, 258
Darlton (Notts.) 13
Darrington (Yorks.) 24
Datchworth (Herts.) 13, 24
Daventry (Northants.) 50, 74, 78, 93, 103, 112, 130, 161, 168, 176, 185, 188, 190, 202, C7, C18, T8B
   electricity 290
   rural district 313
Daventry & Weedon Railway 130, 161, 168, 176
Daventry Railway 78, 93
Davis, John T10
   Richard T10, T12, T13
Deacon: *see* Binnie, Deacon & Gourley
Dean (Beds.) 87
Deanshanger (Northants.), gas 315
   see also Passenham
Deddington (Oxon.) 7
Deene (Northants.) 145
Deenethorpe (Northants.) 145
Deeping: *see also* Market Deeping
Deeping Fen (Lincs.) 13, 57, 19, 27, 35, 49, 59, 100, C14
Deeping Gate (Northants.) 97, 127, 162, C14
Deeping St James (Lincs.) 6, 13, 19, 27, 57, 59, 89, 97, 100, 111, 127, 232, C14, C15
Deeping St Nicholas (Lincs.) 100, 160
Denford (Northants.) 10, 64, 87
Denham (Bucks.) 18, 32, 262, C7
Dennis: *see* Nixon & Dennis
Denton (Hunts.) 13, 24
Denton (Northants.) 25, 64, 104, T7
   electricity 290
   gas 309
Desborough (Northants.) 20, 23, 33, 42, 56, 67, 284
   electricity 289
Diddington (Hunts.) 13, 24
Didsbury (Lancs.) 23
Digby (Lincs.) 6, 16, 49, 97, 127, 160
Digswell (Herts.) 13, 24
Dilhorne (Staffs.) 23
Dingley (Northants.), electricity 289
Dinnington (Yorks.) 135
Direct London & Manchester Railway 23
Direct Northern Railway 13
Direct Northern Railway from London to York 24
   see also London & York Railway
Dockray, Robert B. 53
Dodderhill (Worcs.) 22, 36
Doddington: *see* Dry Doddington; Great Doddington
Dodford (Northants.) 3, 4, 50, 74, 78, 99, 103, 112, 124, 130, 133, 161, 168, 176, 185, 188, 190, 216, C7, T2, T8B
   electricity 290
   gas 305
Dogdyke (Lincs.) 13
Doncaster (Yorks.) 13, 24, 59, 89, 135
Donington (Lincs.) 160
Dormston (Worcs.) 22, 85

INDEX

Dorrington (Lincs.) 6, 16, 23, 49, 97, 127, 160
Dorsington (Warwicks.) 14, 29
Dorton (Bucks.) 262
Doveridge (Derbys.) 23
Dowdeswell (Gloucs.) 138
Dowsby (Lincs.) 6, 16, 49, 60, 89, 97, 127, C15
Draughton (Northants.) 8, 42, 56, 66
  electricity 294
Drax (Yorks.) 13, 24
Draycott in the Clay (Staffs.) 23
Drayton (Leics.) 8, 132, 137, 142
Drayton (Middlesex) C7
Drayton (Oxon.) 92
  see also East Drayton; Old Stratford & Drayton
Drayton Beauchamp (Bucks.) 38
Dringhouses (Yorks.) 13
Droitwich (Worcs.) 22
Dry Doddington (Lincs.) 13
Duckmanton: *see* Sutton cum Duckmanton
Duddington (Northants.) 28, 33, 170, C13, C14
Dudley (Worcs.) 14
Dumbleton (Gloucs.) 47
Dunchurch (Warwicks.) 11, T8B
Dunsby (Lincs.) 6, 16, 49, 60, 89, 97, 127, C15
Dunsden: *see* Eye & Dunsden
Dunstable (Beds.) T8B
Dunston (Lincs.) 6, 13, 16, 49, 97, 127, 160
Dunton Bassett (Leics.) 209, 213
Durham, John T7
  John jun. T15
Duston (Northants.) 1, 8, 29, 31, 39, 42, 50, 56, 64, 66, 99, 103, 104, 124, 129, 133, 134, 149, 152, 155, 171, 179, 185, 190, C4, C7
  electricity 200, 258
Duston Minerals & Northampton & Gayton Junction Railway 155

Ealing (Middlesex) 262
Eardington (Salop) 262
Earith: *see* Bluntisham cum Earith
Earls Barton (Northants.) 10, 64, 122
  electricity 286
  Ferrers & Rushden water 249
Easenhall (Warwicks.) C25
East & West Junction Railway 85, 94, 104, 107, 134, 148
East & West of England Junction Railway 34
East Barnet (Herts.) 13, 23, 24, 52, T8B
East Bridgford (Notts.) 135
East Carlton (Northants.) 19, 28, 33, 37, 132, 137, 142, C13, C14
  electricity 296
  gas 304
East Claydon (Bucks.) 30, 38, 51, 53, 151, 189, 199, 234
East Drayton (Notts.) 13
East Farndon (Northants.) 8, 37, 66, 118, C4
  electricity 289
East Haddon (Northants.) 149, 217
  electricity 290
  gas 305
East Langton (Leics.) 20, 23, 67, C13, C14
East Leake (Notts.) 209, 213
East Markham (Notts.) 13
East Norton (Leics.) 86, 88, 91, 135, 137, 142
Eastcotts (Beds.) 13, 67
Eastern Counties Railway 49, 60, 79
Easton (Hunts.) 42, 56, 87
Easton (Lincs.) 31
  see also Great Easton
Easton Maudit (Northants.) 61, 108, 119
  electricity 288

Easton Neston (Northants.) 34, 51, 55, 84, 169, 187, 199, 234, C24
  electricity 290
  gas 292
Easton Neston Mineral & Towcester, Roade & Olney Junction Railway 169
Easton on the Hill (Northants.) 15, 19, 28, 33, 41, C13, C14, C15
  electricity 297
  gas 312
Eastrop (Hants.) 30
Eatington (Warwicks.) 14, 22, 85, 94, 272
Eaton (Leics.) 132
Eaton (Notts.) 13
Eaton Bray (Beds.) 1
Eaton Socon (Beds.) 24, 29, 39
Ebrington (Gloucs.) 47
Eckington (Northants.) C10, C12
Ecton (Northants.) 29, 39, 64, 122
  electricity 286
  Higham Ferrers water 237
Edenham (Lincs.) 127
Edgcote (Northants.) 34, C24
Edgcott (Bucks.) 14
Edlesborough (Bucks.) 1
Edlington (Lincs.) 13
Edmondthorpe (Leics.) 15, 40, 41, 58
Edmonton (Middlesex) 13, 23, 52
Edmundson's Electricity Corporation Ltd 297
Edwalton (Notts.) 132
Edwinstowe (Notts.) 135
Eggbrough (Yorks.) 6
Egleton (Rutland) 31, C15
Eldersfield (Worcs.) 125
Electrical Power Distribution Co. Ltd 240
Ellington (Hunts.) 33, 42, 56
Elliott-Cooper, R. 234
  Sir R. & Son 300, 302
  see also Cooper, R. Elliott
Elm (Cambs.) 45, 64, 100
Elmbridge (Worcs.) 36
Elmdon (Warwicks.) 1, T8B
Elmfield: *see* Carr House & Elmfield
Elmsall: *see* North Elmsall; South Elmsall
Elstow (Beds.) 20, 23, 42, 67
Elton (Hunts.) 10, 41, 64, 75, 141
Elton (Notts.) 137, 142
Elwes: *see* Wells-Owen & Elwes
Emberton (Bucks.) 61, 119
  gas 309
Emneth (Norfolk) 64, 100
Empingham (Rutland) 40
Enfield (Middlesex) 13, 24, 52
Epperstone (Notts.) 135
Epwell (Oxon.) 92
Epworth (Lincs.) 13, 97, 160
Essendine (Rutland) 13, 24, 52, 65, 69, 219
Ettington: *see* Eatington
Etton (Northants.) 12, 13, 15, 17, 19, 49, 52, 60, 97, 150, 156, 162, 208
Eunson, John 114
Evans, E.A. 260
Evedon (Lincs.) 97
Evenley (Northants.) 18, 38, 196, 203, 209, 213
Everard, Son & Pick 273
Everdon (Northants.) 22, 29, 36
  electricity 290
Eversden: *see* Great Eversden; Little Eversden
Everton (Beds.) 13, 24
Everton (Notts.) 13
Evesham (Worcs.) 14, 85
Evetts, Geo. 304
Ewerby (Lincs.) 6, 16, 49
Eydon (Northants.) 85, 94, 209, 213, 218, 226
Eye & Dunsden (Oxon.) 51
Eye (Northants.) 6, 19, 35, 44, 46, 54, 59, 79, 81, 100, 160, 276, C14, C15
Eynesbury (Hunts.) 13, 29, 39

Fairbank, W.F. 13
Falcutt: *see* Astwell with Falcutt
Far Cotton (Northants.) 64, 66
  see also Northampton
Farnborough (Warwicks.) 7, 11, 14, 94
Farndon: *see* East Farndon
Farnham Royal (Bucks.) 32, 262
Farnsfield (Notts.) 135
Farrar, T.S. 111
Farthinghoe (Northants.) 21, 38, 47, 53, 55, 84, 120
Farthingstone (Northants.), electricity 294
Fawley (Bucks.) 30
Fawsley (Northants.) 22
  electricity 294
Faxton (Northants.), electricity 294
Featherstone (Yorks.) 24
Feckenham (Worcs.) 36
Felmersham (Beds.) 20, 23, 42, 56, 67, 136, 154
Fen Drayton (Cambs.) 16, 49
Fen Stanton (Cambs.) 16, 49
Fenny Compton (Warwicks.) 11, 14, 22, 85, 94
Fenny Stratford (Bucks.) 61, C7, T8B
Fenton (Lincs.) 24
  see also Pidley cum Fenton
Fenwick (Yorks.) 6, 13
Ferry Frystone (Yorks.) 24
Feuvre, Le: *see* Ordish & Le Feuvre
Fidler, Alfred 271, 274
Filgrave: *see* Tyringham with Filgrave
Fillingham (Lincs.) 160
Finchley (Middlesex) T8B
Finedon (Northants.) 20, 33, 42, 56, 63, 67, 136, 175, 215
  electricity 288
  gas 257
  Nene Valley water 231
  tramroads 239
  water 248
Finedon Gas Co. Ltd 257
Fineshade (Northants.) 141
Finmere (Oxon.) 38, 196, 203, 209, 213
Finningley (Notts.) 6, 24, 59, 89
Fishlake (Yorks.) 6, 13, 24, 97, 160
Fishtoft (Lincs.) 13
Fiskerton (Lincs.) 13, 31
Fladbury (Worcs.) 14
Flamstead (Herts.) T8B
Fleckney (Leics.) C4
Fledborough (Notts.) 13
Fleet Marston (Bucks.) 14, 18, 38, 196, 203
Fletton (Hunts.) 10, 13, 15, 24, 35, 44, 45, 46, 49, 52, 59, 62, 64, 72, 97, 115, 139, 269, 275
Flitton (Beds.) 23, 107, 128
Flitwick (Beds.) 107, 128
Flore (Northants.) 1, 50, 99, 103, 124, 133, 185, 190, T2
  electricity 290
  gas 292
Floyd, Thomas 171, 179
Flyford Flavel (Worcs.) 85
Foleshill (Warwicks.) C21, C25
Folkingham (Lincs.) 60, 97, 127
Folksworth (Hunts.) 13, 24
Forest Hill with Shotover (Oxon.) 32
Forsbrook (Staffs.) 23
Forthampton (Gloucs.) 125
Foscott (Bucks.) 30, 51, C5, T4
Fosdyke (Lincs.) 13
Fotheringhay (Northants.) 10, 64
Foulby: *see* Huntwick with Foulby & Nostell
Fowler, Sir John Bt 72, 94a, 98, 115, 146,

183, 204, 208, 219, 225
see also Hawkshaw, Fowler & Stephenson
Fox, Sir Charles & Son 103
  Sir Douglas & Francis 218, 221, 223
  Sir Douglas & Frank 226
  Sir Douglas, Francis & Douglas 228, 233
  see also Martin, Johnson & Fox
Foxton (Leics.) 8, 20, 23, 42, 56, 67, 302, C4, C10, C11, C12, C13, C14
Frampton (Lincs.) 13, 19, 49
Frankton (Warwicks.) 7
Fraser, John 137, 142, 146, 158
  John & Sons 204, 208
Fraser & Stanton 89
Freebody, William Yates 78
Freeby (Leics.) 40, 86, 91, 132
Friern Barnet (Middlesex) 13, 23, 24, 52, T8B
Frisby on the Wreak (Leics.) 15, 41, 137
Frithville (Lincs.) 13
Fritwell (Oxon.) 11, 14, 32, 262
Fulbeck (Lincs.) 13, 31
Fulbrook (Warwicks.) 36
Fulham (Middlesex) 3
Fulmer (Bucks.) 32, 262
Fulton, Hamilton C15
Furtho (Northants.) 107, 128, C7, T8B
  gas 309

Gaddesby (Leics.) 86, 91, 137
Gaddesden: *see* Great Gaddesden; Little Gaddesden
Gainsborough (Lincs.) 13, 89, 97
Gamston (Notts.) 40, 43
Garthorpe (Leics.) 40
Gate Burton (Lincs.) 13
Gateforth (Yorks.) 6
Gawsworth (Cheshire) 23
Gaydon (Warwicks.) 94
Gayton (Northants.) 1, 3, 10, 21, 47, 55, 84, 85, 95, 104, 129, 134, 148, 152, 155, 189, 224, 317, C7
  electricity 290
  gas 292
Geddington (Northants.) 31, 43, 145
  electricity 289
  gas 304
General Gas & Electricity Co. Ltd 313
Gibbins, W.D. 242
Giles, Francis 7, 8, 9
Gilmorton (Leics.) 209, 213
Gilroes (Leics.) 209
Girton (Cambs.) 6, 16, 49, 127
Glapthorn (Northants.) 10, 64
  electricity 291
Glaston (Rutland) 132, 145
Glatton (Hunts.) 13, 24
Glen Magna (Leics.) 23, 56, C4
Glen Parva (Leics.) C4
Glendon (Northants.) 42, 56, 67, 136, 145, 165
  gas 304
Glenfield (Leics.) 23, 209, 213
Glentworth (Lincs.) 160
Glinton (Northants.) 12, 13, 15, 16, 17, 19, 49, 52, 89, 97, 100, 111, 127, 150, 156, 162, 208, C14, C15
Glooston (Leics.) 137, 142
Goadby (Leics.) 137, 142
Goadby Marwood (Leics.) 132
Godington (Oxon.) 203, 209, 213
Godmanchester (Hunts.) 13, 42, 56, 87
Goldington (Beds.) 13, 67
Gonalston (Notts.) 135
Gonerby: *see* Great Gonerby
Gordon: *see* Liddell & Gordon
Gosberton (Lincs.) 13, 19, 49, 160
Gotham (Notts.) 209, 213
Gottschalk, Julius 214

Gourley: *see* Binnie, Deacon & Gourley
Grafham (Hunts.) 87
Grafton Flyford (Worcs.) 22, 85
Grafton Manor (Worcs.) 36
Grafton Regis (Northants.) C7
  gas 309
Grafton Underwood (Northants.) 191
  electricity 296
  gas 304
Granby (Notts.) 137, 142
Grand Junction Canal C7, C9, C18
  Buckingham Branch C5
Grand Union Canal 300, 302
  see also Old Union Canal
Grandborough (Bucks.) 38, 53
Grandborough (Warwicks.) 14, 36, 50, 74, 93, 103, 190, 202, 300, C6, T1
Gransden: *see* Great Gransden
Grantchester (Cambs.) 6, 29, 39, 127
Grantham (Lincs.) 13, 24, 31, 52
Grassthorpe (Notts.) 13
Gravatt, W. 13
Graveley (Herts.) 13, 24
Grazeley (Berks.) 30
Great Addington (Northants.) 10, 33, 42, 56, 64, 80, 175
  electricity 291
  gas 307
Great Alne (Warwicks.) 36
Great & Little Hampton (Worcs.) 14, 85
Great Billing (Northants.) 29, 39, 64, 122
  electricity 258
Great Bowden (Leics.) 8, 19, 20, 23, 28, 33, 37, 42, 56, 67, 157, 173, 184, C4, C11, C13, C14
Great Brickhill (Bucks.) 1, T8B
Great Casterton (Rutland) 40
Great Central Railway 228, 233
Great Corringham: *see* Corringham, Great & Little
Great Creaton (Northants.) 42, 66, C4
  electricity 290
  gas 292
Great Dalby (Leics.) 86, 91, 135, 137
Great Doddington (Northants.) 10, 61, 64, 108
  electricity 288
  Great Ouse water 318
  Higham Ferrers & Rushden water 249
Great Eastern Northern Junction Railway 89
Great Eastern Railway 160
Great Easton (Leics.) 8, 28, 31, 33, 37, 132, 137, 142
  Corby water 299
Great Eversden (Cambs.) 29, 39
Great Gaddesden (Herts.) 1
Great Glen: *see* Glen Magna
Great Gonerby (Lincs.) 13, 24, 31, 52
Great Gransden (Hunts.) 29, 39
Great Hale (Lincs.) 6, 16, 49
Great Harrowden (Northants.) 20, 23, 42, 43, 56, 63, 67, 136, 215
  electricity 288
  gas 307
Great Houghton (Northants.) 10, 25, 64, 99, 103, 104, 106, T7
  electricity 257
Great Kimble (Bucks.) 51
Great Linford (Bucks.) 107, 128, C7, T4
Great Marlow (Bucks.) 51
Great Milton (Oxon.) 32
Great Missenden (Bucks.) 18
Great Northern Railway, bridge over Welland at Deeping 232
  deviations between London and Grantham 52
  deviations between Peterborough, Doncaster and Boston 59
  footpath at Lolham Bridges 225
  footpath at Peterborough 279

land at Fletton 269
land at Lolham Bridges 268
land at New England 158, 219
land at Peakirk 255
land at Peterborough 139, 251, 255, 279
land at Peterborough and Helpston 146
land at Walton 278
land at Werrington 236
land near Werrington Junction 204
new line at New England 236
new line at Peterborough 219
new line at Werrington Junction 219
subway at Peterborough 183
widening at Peterborough Station 232
widening from Helpston to Essendine 219
widening from Werrington Junction to Helpston 208
widenings at Peterborough 275
see also London & York Railway
Great Northern and London & North Western Railways 142
Great Oakley (Northants.) 31, 43, 132, 145
  electricity 296
  gas 304
Great Ouse, water 318
Great Oxendon (Northants.) 8, 28, 33, 42, 56, 66, 141, 144, 224, C4
  electricity 296
Great Packington: *see* Packington, Great & Little
Great Paxton (Hunts.) 13
Great Ponton (Lincs.) 13, 24, 31, 52
Great Rollright (Oxon.) 34, 94a, 98, 101, 102, 105, 138
Great Stukeley (Hunts.) 13, 33, 42, 56
Great Weldon (Northants.) 145
  electricity 296
  gas 304
  see also Weldon
Great Western Railway 126, 261, 262, 270
Great Wolford (Warwicks.) 92, 102, 125, 140
Great Woolstone (Bucks.) 61
Great Wymondley (Herts.) 13, 24
Greatford (Lincs.) 13, 49, 150, 156, 219, C15
Greatworth (Northants.) 21, 38, 47, 53, 55, 84, 120
Greens Norton (Northants.) 84, 85, 94, 151, C24, T8B
  electricity 290
  gas 298
Greetham (Rutland) 40
Greetwell (Lincs.) 13, 31, 89, 97, 127
Gregory, C.H. 43
Grendon (Northants.) 29, 39, 61, 64, 108, 119
  electricity 288
  gas 312
Grendon Underwood (Bucks.) 14, 30, 196, 203, 209, 213
Gretton (Northants.) 28, 33, 43, 145, 157, C13, C14
  electricity 296
Grierson, W.W. 261, 262, 270
Grimsbury (Northants.) 34, 53, 55, 68, 85, 92, 95, 96, 101, 102, 105, 125, 140
Grimston (Leics.) 132
Gringley on the Hill (Notts.) 13, 24, 59
Grove (Bucks.) 3, C7
Guilsborough (Northants.) 186
  electricity 290
  gas 305
  water 281
Guiting Power (Gloucs.) 138
Gumley (Leics.) 302, C4, C10, C11
Gunthorpe (Northants.) 79, 81, 97, 127, C14, C15
Gunthorpe (Notts.) 135, C15
Gunthorpe (Rutland) 31

INDEX

Guyhirn (Cambs.) 64, 79, 81

Habblesthorpe: *see* North Leverton with Habblesthorpe
Hacconby (Lincs.) 6, 16, 49, 60, 89, 97, 127, C15
Hackleton (Northants.) 99
   electricity 290
   gas 292
Hackney (Middlesex) 127
Haddenham (Bucks.) 30
Haddlesey: *see* Chapel Haddlesey
Haddon (Hunts.) 13, 24
   see also East Haddon; West Haddon
Hadley (Herts.) T8B
Hale: *see* Great Hale; Little Hale
Hall: *see* Sherrard & Hall 52
Hallaton (Leics.) 86, 88, 91, 135, 137, 142
Halstead (Leics.) 86, 91, 135, 137
Hambleton (Rutland) 31
Hambleton (Yorks.) 6
Hammersmith (Middlesex) 3, 262
Hamphall Stubbs (Yorks.) 13
Hampole (Yorks.) 13
Hampstead (Middlesex) 209, 213
Hampton: *see* Great & Little Hampton
Hampton in Arden (Warwicks.) 1, 3, C21, T8B
Hampton Gay (Oxon.) 7, 11
Hampton Poyle (Oxon.) 7
Hanbury (Staffs.) 23
Hanbury (Worcs.) 36
Handforth (Cheshire) 23
Hanging Houghton (Northants.) 66, C4
   electricity 290
   gas 292
Hannington (Northants.), 31, T5
   electricity 290
   gas 292
   Great Ouse water 318
Hanslope (Bucks.) 1, 3, 149
   gas 309
Hanwell (Middlesex) C7
Hanwell (Oxon.) 85, T3
Harbury (Warwicks.) 29, 50, 103
Harby (Leics.) 137, 142
Hardingstone (Northants.) 1, 8, 10, 25, 29, 31, 39, 42, 48, 50, 56, 64, 66, 77, 82, 83, 99, 103, 104, 106, 110, 147, 149, 152, 171, C7, T7
   electricity 200, 258
Hardwick (Lincs.) 13
Hardwick (Northants.), electricity 288
   gas 307
   Great Ouse water 318
Hardwick (Oxon.) 85, 95, 101, 105
Harefield (Middlesex) C7
Hargrave (Northants.) 87
   electricity 291
Harlestone (Northants.) 149, 167, 181, 216
   electricity 290
Harlington (Middlesex) C7
Harlton (Cambs.) 29, 39
Harmston (Lincs.) 13, 31, 89
Harpenden (Herts.) 23, T8B
Harpole (Northants.) 50, 103, 133, 185
   electricity 290
Harpsden (Oxon.) 30, 51
Harpswell (Lincs.) 160
Harrington (Northants.) 33, 66
   electricity 296
   gas 304
Harringworth (Northants.) 28, 33, 145, C11, C13, C14
Harris, William Cecil 293
Harrold (Beds.), electricity 287
   gas 309
Harrow (Middlesex) 1, 3
Harrowby (Lincs.) 13, 31
Harrowden: *see* Great Harrowden; Little Harrowden
Hartlebury (Worcs.) 14
Hartpury (Gloucs.) 125
Hartwell (Bucks.) 196, 203
Hartwell (Northants.) 1, 3, 149
   gas 309
Harvington (Worcs.) 85
Harworth (Notts.) 13
Haselbech (Northants.), electricity 294
Haseley (Warwicks.) C24
Haslingfield (Cambs.) 29, 39
Hatfield (Herts.) 13, 24, 52
Hatfield (Yorks.) 6, 13, 24, 97, 160
Hatton (Warwicks.) 300, C24
Haverholme Priory (Lincs.) 49
Haversham (Bucks.) 1, 3, 4, 107, 128, 149
Hawkesley, Thomas 43
   see also Hawksley
Hawkshaw, Fowler & Stephenson 89
Hawkshaw, Sir John 97, 160
Hawksley, T. 96
   T. & C. 186, 316
   see also Hawkesley
Haxey (Lincs.) 6, 13, 24, 59, 89, 97, 160
Hayes (Middlesex) 32, C7
Haynes (Beds.) 20, 24
Hayward, F.A. 117
   R. & F. 74
Headham cum Upton (Notts.) 13
Headington (Oxon.) 32
Heapham (Lincs.) 89, 97
Heath: *see* Warmfield cum Heath
Heaton Norris (Lancs.) 23
Heck (Yorks.) 6, 13
Heckington (Lincs.) 6, 16, 49
Hedgerley (Bucks.) 32
Hedgerley Dean (Bucks.) 32
Heighington (Lincs.) 6, 13, 16, 49, 97, 127
Hellaby: *see* Stainton with Hellaby
Hellidon (Northants.) 29, 134, 209, 213
   electricity 294
Helmdon (Northants.) 21, 34, 47, 55, 84, 95, 120, 196, 203, 209, 213, 223, C24
Helpringham (Lincs.) 6, 16, 49, 160
Helpston (Northants.) 12, 13, 15, 17, 19, 52, 97, 146, 162, 208, 219, 225, 268
Hemans, G.W. 93
Hemel Hempstead (Herts.) 1, 3, C7
Hemington (Northants.), electricity 291
Hemswell (Lincs.) 160
Hemsworth (Yorks.) 13
Henley on Thames (Oxon.) 30
Henlow (Beds.) 13, 24, 42, 67
Hennell, Thomas 131, 191, 194, 195, 237
Hensall (Yorks.) 6, 13
Heston & Isleworth (Middlesex) 300, C7
Hexthorpe: *see* Balby with Hexthorpe
Heyford: *see* Lower Heyford; Nether Heyford; Upper Heyford
Heythrop (Oxon.) 34, 94a, 102
Hickling (Notts.) 40, 132
High Wycombe: *see* Chepping Wycombe
Higham Ferrers (Northants.) 10, 64, 201, 211, 212
   electricity 259, 264, 277, 287
   gas 307
   tramroads 198, 205, 239, 260
   water 237
Higham Ferrers & Rushden Water Board 249
Higham Ferrers Water Co. 237
Higham Gobion (Beds.) 23, 107, 128
Highworth (Wilts.), rural district 313
Hillingdon (Middlesex) 32, 262, C7
Hillmorton (Warwicks.) 1, 3, 4, 5, 11, 14, 149, 209, 213, 218, C21, C25
Himbleton (Worcs.) 22
Himley (Staffs.) 262
Hinksey: *see* North Hinksey; South Hinksey
Hinton in the Hedges (Northants.) 38, 94

Hirst, William 69
Histon (Cambs.) 6, 16, 49, 127
Hitchin (Herts.) 13, 20, 24, 42, 67, 107
Hoby (Leics.) 15, 41
Hockcliffe (Beds.) T8B
Hodnell: *see* Upper Hodnell
Hodsock (Notts.) 135
Hogshaw (Bucks.) 38, 51
Holcot (Beds.) 107, 128
Holcot (Northants.), electricity 290
   gas 292
   Northampton water 316
Holdenby (Northants.) 149, 181, 186
   electricity 290
   gas 305
   Northampton water 281
Holdingham (Lincs.) 89, 97, 127
Hollowell (Northants.), electricity 290
   gas 305
   Northampton water 281, 282
   water 281
Holme (Hunts.) 13, 97
Holme Pierrepont (Notts.) 40, 43
Holt (Leics.) 132, 137, 142
Holt (Norfolk) 279
Holton (Oxon.) 32
Holwell (Herts.) 13, 20, 24, 42, 67
Holwell (Leics.) 40, 135
Holywell (Lincs.) 52
Holywell cum Needingworth (Hunts.) 16, 49
Honington (Lincs.) 13
Hook Norton (Oxon.) 34, 47, 68, 92, 98, 101, 102, 105, 125, 138, 140
Hooton Levitt (Yorks.) 13
Hooton Pagnell (Yorks.) 13
Hopcraft, Thomas T1
Hopkins, George 152
Horbling (Lincs.) 6, 16, 49, 89, 97, 127
Horninghold (Leics.) 86, 88, 91, 137, 142
Hornsey (Middlesex) 13, 23, 24, 52, T8B
Horsendon (Bucks.) 32, 51
Horsington (Lincs.) 13
Horton (Northants.) 99, 103, 169
   electricity 290
   gas 292
Hose (Leics.) 43, 135
Hothorpe (Northants.) C4, C13, C14
   electricity 296
Hough on the Hill (Lincs.) 24
Hougham (Lincs.) 13, 24, 31
Houghton (Hunts.) 16
   see also Great Houghton; Hanging Houghton; Little Houghton; Wyton cum Houghton
Houghton Conquest (Beds.) 23
Houghton Regis (Beds.) T8B
Houghton, Dugdale C21, C24
Hucknall Torkard (Notts.) 26, 209, 213
Huddington (Worcs.) 22
Hughes, Samuel 59
Hulcott (Bucks.) 14
Hulme (Lancs.) 23
Hunningham (Warwicks.) 202
Huntingdon (Hunts.) 13, 24, 33, 42, 56, 87
Huntley (Staffs.) 23
Huntwick with Foulby & Nostell (Yorks.) 13
Hurley (Berks.) 51
Hurst, William 65
Husbands Bosworth (Leics.) 33, 37, C10, C12, C13, C14
   Rugby water 285
Husborne Crawley (Beds.) 107, 128

Ibstock (Leics.) 23
Ickford (Bucks.) 32
Ickleford (Herts.) 13, 20, 24, 42, 67, 107, 128
Iffley (Oxon.) 32
Impington (Cambs.) 6, 16, 49, 127

Ingham (Lincs.) 160
Ingleby: *see* Saxilby cum Ingleby
Inglis, James C. 261, 262, 270
   R.J.M. 311
Inkberrow (Worcs.) 22, 85
Ippollitts (Herts.) 13, 24
Ipsley (Worcs.) 36, 85
Ipstones (Staffs.) 23
Irchester (Northants.) 10, 20, 23, 42, 43, 56, 61, 63, 64, 67, 108, 136, 154, 182, 197, 201, 212
   electricity 287, 288
   Great Ouse water 318
   tramroads 198, 205, 239, 260
Irthlingborough (Northants.) 10, 20, 23, 42, 43, 56, 63, 64, 67, 136, 154, 253
   electricity 259, 277, 288
   gas 238, 307
   tramroads 239
Irthlingborough Gas & Coke Co. Ltd 238, 307
Isham (Northants.) 20, 23, 42, 43, 56, 63, 67, 83, 136, 174, 193, 314
   electricity 288, 289
   gas 304
Isleworth: *see* Heston & Isleworth
Islington (Middlesex) 13, 23, 24, 52, T8B
   see also Tilney with Islington
Islip (Northants.) 10, 64, 80, 87, 175, T14
   electricity 291
Islip (Oxon.) 14
Itchington: *see* Bishops Itchington; Long Itchington
Iver (Bucks.) 18, 32, 262
Ivinghoe (Bucks.) 3, 14, C7

Jackson, G.H. 277, 291, 294
Johnson, Richard 139, 146, 158, 183, 204, 208, 219, 225
   see also Martin, Johnson & Fox
Johnston, Robert E. 126
Johnstone, A.D. 112, 124, 130, 133, 161, 168

Kelham (Notts.) 24
Kellington (Yorks.) 6
Kelmarsh (Northants.) 8, 33, 42, 56, 66, 141, 144, C4
   electricity 296
Kempston (Beds.) 23, 25, 103
Kensington (Middlesex) 3
Kensworth (Beds.) T8B
Kershaw, G. Bertram 285
Kettering (Northants.) 20, 23, 31, 33, 42, 43, 56, 67, 80, 123, 136, 145, 157, 164, 230, 254, T5
   Burton Latimer water 273
   electricity 222, 289, 295, 296
   gas 177, 304, 314
   street improvements 256
   water 131, 191, 194, 195, 243
Kettering & Thrapstone Railway 80, 87
Kettering Gas Co. 304, 314
Kettering Gas Co. Ltd 177
Kettering Urban District Council 222, 256, 273, 289, 296
Kettering Urban District Water 243
Kettering Waterworks Co. Ltd 194
Kettleby: *see* Ab Kettleby
Kettlethorpe (Lincs.) 24
Ketton (Rutland) 8, 15, 19, 28, 33, C15
   electricity 297
Kexby (Lincs.) 97
Keysoe (Beds.) 29, 39
Keyston (Hunts.) 33, 42, 56, 87
Kibworth Beauchamp (Leics.) 8, 19, 20, 23, 42, 56, 67, C4
Kibworth Harcourt (Leics.) 8, 20, 67, C4, C13, C14
Kidderminster (Worcs.) 14, 36

Kidlington (Oxon.) 7, 11, 14, 53
Kilsby (Northants.) 1, 3, 4, 5, 149, C12, C25, T2
Kilworth: *see* North Kilworth; South Kilworth
Kimble: *see* Great Kimble
Kimbolton (Hunts.) 87
Kineton (Warwicks.) 14, 22, 85, 94, 272
King, John T6
Kingham (Oxon.) 34, 98, 138
Kings Cliffe (Northants.) 141, 159, 163, 170
Kings Langley (Herts.) 3, C7
Kings Newnham (Warwicks.) C21, C25
Kings Sutton (Northants.) 7, 11, 14, 18, 21, 38, 47, 92, 94a, 126, 138, 261, 313
Kingsbury (Middlesex) 1
Kingsley (Staffs.) 23
Kingsthorpe (Northants.) 8, 31, 42, 56, 64, 66, 149, 175, 179, 181, 186, T5
   electricity 200
   Northampton water 281
   see also Northampton
Kingston: *see* Chesterton & Kingston
Kingston (Cambs.) 29, 39
Kingswinford (Staffs.) 14, 262
Kington (Worcs.) 22
Kinoulton (Notts.) 40
Kinver (Staffs.) 14
Kirby Bellars (Leics.) 15, 86, 91, 137
Kirby Frith (Leics.) 23
Kirby Muxloe (Leics.) 23
Kirk Bramwith (Yorks.) 6, 97
Kirk Sandall (Yorks.) 89, 97
Kirk Smeaton (Yorks.) 24
Kirkby Green (Lincs.) 97, 127, 160
Kirkby in Ashfield (Notts.) 26, 209, 213
Kirkby la Thorpe (Lincs.) 49, 97, 127, 160
Kirkby Underwood (Lincs.) 127
   see also South Kirkby
Kirkstead (Lincs.) 13
Kirtlington (Oxon.) 7, 11
Kirton (Lincs.) 13, 19, 49
Kislingbury (Northants.) 1, 29, 39, 48, 50, 64, 99, 103, 104, 124, 129, 133, 134, 152, 155, 185, 190
   electricity 258
Knaith (Lincs.) 13
Knebworth (Herts.) 13, 24
Knighton (Leics.) 23
Knotting (Beds.) 29, 39
   electricity 287
Knowle (Warwicks.) 300

Ladbroke (Warwicks.) 7, 11, 14, 29, C24
Lamport (Northants.) 8, 42, 56, 66, 73, C4
   electricity 290
   gas 292
Lancashire & Yorkshire and Great Eastern Junction Railway 97
Lane, John 85
Laneham (Notts.) 24
Langar cum Barnston (Notts.) 135, 137, 142
Langford (Beds.) 13, 24
Langham (Rutland) 43
Langley (Warwicks.) 36
   see also Abbots Langley; Kings Langley
Langley Marish (Bucks.) 32, 262
Langley, Alfred A. 201
Langriville (Lincs.) 13
Langtoft (Lincs.) 6, 16, 49, 60, 89, 97, 127, 150, 156
Langthwaite with Tilts (Yorks.) 13
Langton: *see* Church Langton; East Langton; Thorpe Langton; Tur Langton; West Langton
Langworth: *see* Stainton by Langworth
Lapworth, Herbert 299
Lapworth (Warwicks.) C24
Lathbury (Bucks.) 61
Laughton (Lincs.) 60, 97, 127, 160

Laughton en le Morthen (Yorks.) 13, 135
Launde (Leics.) 86, 91
Launton (Oxon.) 14, 262
Lavendon (Bucks.) 103, 104, 108, 110, 119, T7
   gas 309
   tramroads 205
Law, E.F. 120a
Lawford: *see* Church Lawford; Little Lawford; Long Lawford
Lawson & Mansergh 122
Laxton, W. 44
Laxton (Northants.) 145
Le Feuvre: *see* Ordish & Le Feuvre
Lea (Lincs.) 13, 97
Leadenham (Lincs.) 13, 31
Leake: *see* East Leake
Leamington Hastings (Warwicks.) 11, 50, 202, C6, T1
Leamington Priors (Warwicks.) 36, 48, 50, 103, 190
Leane & Bakewell 190
   see also Cooper, R. Elliott, Leane & Bakewell
Leasingham (Lincs.) 89, 97, 127, 160
Leckhampstead (Bucks.) 30, 51, C5, T4
Leckhampton (Gloucs.) 138
Leek (Staffs.) 23
Leek Wootton (Warwicks.) C24
Leekfrith (Staffs.) 23
Leesthorpe: *see* Pickwell with Leesthorpe
Leicester 8, 20, 23, 42, 209, 213, C4
Leicester & Bedford Railway 20, 28
Leicester, Northampton, Bedford & Huntingdon Railway 42
Leicestershire & Northamptonshire Union Canal C2, C3, C4
Leigh (Staffs.) 23
Leighton Bromsworld (Hunts.) 33, 42, 56
Leighton Buzzard (Beds.) 1, C7
Lemington: *see* Lower Lemington
Lenchwick (Worcs.) 85
Lenton (Notts.) 26
Letchworth (Herts.) 13, 24
Leverington (Cambs.) 19, 46, 62, 64, 79, 81
Leverton: *see* North Leverton; South Leverton
Lewin, William 40
Lewknor (Oxon.) 30
Leyton (Essex) 127
Liddell, Charles 15, 37, 41, 42, 56, 67, 92, 99, 106, 120, 125, 129, 140, 187, 196, 203, 209, 213
Liddell & Gordon 68
Liddington (Rutland) 8, 19, 28, 31, 33, 37, 43, 132, 145
Lidlington (Beds.) 107, 128
Lilbourne (Northants.) 33, 37, 167
   Rugby water 285
Lilburne, T. T5
Lilford cum Wigsthorpe (Northants.) 10, 64
   electricity 291
Lillingstone Dayrell (Bucks.) T6
Lillingstone Lovell (Bucks.) 51, T6
Lillington (Warwicks.) C24
Limbury (Beds.) 23
Linby (Notts.) 26, 209, 213
Lincoln (Lincs.) 13, 16, 24, 31, 49, 89, 97, 160
Linslade (Bucks.) 1, 3, C7
Linton (Herefs.) 125
Linwood (Lincs.) 13, 16, 31, 49
Lissington (Lincs.) 31
Lister & Mills 49
Litchborough (Northants.), electricity 294
Little Addington (Northants.) 42, 56, 64
   electricity 291
   gas 307
   Nene Valley water 231
Little Barford (Beds.) 13, 24

Little Billing (Northants.) 29, 39, 64, 122
  electricity 258
Little Bowden (Northants.) 8, 20, 23, 28, 33, 37, 42, 56, 66, 67, 157, 167, 184
Little Brickhill (Bucks.) T8B
Little Bytham (Lincs.) 13, 24, 52, 275
Little Casterton (Rutland) 40
Little Corringham: *see* Corringham, Great & Little
Little Eversden (Cambs.) 29, 39
Little Gaddesden (Herts.) 1
Little Hale (Lincs.) 6, 16, 49
Little Hampton: *see* Great & Little Hampton
Little Harrowden (Northants.) 20, 23, 42, 43, 56, 63, 67, 136
  electricity 288
  gas 307
  Great Ouse water 318
Little Houghton (Northants.) 10, 25, 64, 99, 104, 224, T7
  electricity 258
Little Lawford (Warwicks.) C25
Little Linford (Bucks.) 107, 128
Little Missenden (Bucks.) 18
Little Oakley (Northants.) 43, 132, 145
  electricity 296
  gas 304
Little Packington: *see* Packington, Great & Little
Little Paxton (Hunts.) 13, 24
Little Ponton (Lincs.) 13, 24, 31, 52
Little Rollright (Oxon.) 94a, 102
Little Stukeley (Hunts.) 33, 42, 56
Little Tew (Oxon.) 94a
Little Weldon (Northants.)
  electricity 296
  gas 304
  see also Weldon
Little Wolford (Warwicks.) 92, 125, 140
Little Woolstone (Bucks.) 61, C7
Little Wymondley (Herts.) 13, 24
Littleborough (Notts.) 6
Littleton: *see* North & Middle Littleton; South Littleton
Littlewood: *see* Alexander & Littlewood
Loddington (Leics.) 86, 91, 135, 137, 142
Loddington (Northants.) 195
  electricity 289
  gas 304
  Kettering water 243
Loddington Ironstone Co. Ltd 220
London & Birmingham Extension, Northampton, Daventry, Leamington & Warwick Railway 50
London & Birmingham Junction Canal C21, C24
London & Birmingham Railway 1, 2, 3, 4, 5, 10, 48
  Banbury Line 21
London & North Eastern Railway 301, 311
London & North Western Railway, Addington Branch 175
  additional land near Castle Station, Northampton 167
  Bletchley, Northampton and Rugby Railway 149
  branch from Northampton to Market Harborough 66
  bridle road at Kings Cliffe 163
  bridle road at Long Buckby 163
  Daventry and Leamington Railway 202
  deviation at Braunston 216
  deviation at Bridge Street, Northampton 77, 82
  diversion of bridle road at Harlestone 167
  diversion of bridle road at Roade Station 167
  diversion of Nene at Northampton 163
  diversion of road at Higham Ferrers Station 211
  diversion of road at Lamport 73
  Islip Branch 175
  junction at Weedon 216
  land at Castle Station, Northampton 235
  land at Clipston & Oxendon Station 224
  land at Courteenhall 229
  land at Duddington 170
  land at East Haddon, Brington and Long Buckby 217
  land at Gayton 224
  land at Hardingstone 147
  land at Harlestone 181, 216
  land at Holdenby 181
  land at Kelmarsh Station 245, 246
  land at Kings Cliffe 170
  land at Kingsthorpe 181
  land at Little Houghton 224
  land at Long Buckby 184
  land at Milton and Wootton 216
  land at Nether Heyford 184, 245, 246
  land at Pitsford 153
  land at Roade 216, 224
  land at Thorpe Achurch 153
  land at Thorpe Station 202
  land at Wellingborough Station 197
  land at Weston 153
  land and extension of bridge at Little Bowden and Great Bowden 184
  new bridge at Milton 245, 246
  new footpath at Brington 167
  new footpath at Catthorpe and Lilbourne 167
  new footpath at Long Buckby 167
  new footpath on Seaton and Wansford line at Kings Cliffe 159
  new railway at Kelmarsh 141, 144
  new railway at Oxendon 141, 144
  new road at Little Bowden 167
  new road at Wakerley 147
  new roads at Thorpe Lubenham, Lubenham and East Farndon 118
  Newport Pagnell, Olney Branch 61
  Northampton and Kingsthorpe widening 175
  Seaton and Wansford Railway 141
  sidings at Hardingstone 77
  Weedon & Daventry Railway 188
  Weedon Deviation 188
  works at Northampton and Pitsford 70
London & North Western Railway and Midland Railway, Market Harborough New Line 173
London & Nottingham Railway 43
London & Worcester & Rugby & Oxford Railway 14
London & York (Great Northern) Railway, Stamford & Spalding Branch 27
London & York Railway 12, 17
  see also Direct Northern Railway
London, Birmingham & Buckinghamhire Railway 18
London, Warwick, Leamington & Kidderminster Railway 36
Long Buckby (Northants.) 1, 3, 4, 149, 163, 167, 184, 217, C7, T2
  electricity 290
  gas 305
Long Buckby Gas-Light Coke & Coal Co. Ltd 305
Long Clawson (Leics.) 43, 135
Long Compton (Warwicks.) 92, 102, 125, 140
Long Crendon (Bucks.) 30, 32
Long Itchington (Warwicks.) 7, 36, 50, 74, 93, 103, 190, 202, 300, C6, C24
Long Lawford (Warwicks.) 1, 3
Long Marston (Warwicks.) 29
Long Stanton All Saints (Cambs.) 6, 49, 89, 97, 127, 160
Long Stanton St Michael (Cambs.) 6, 16, 49, 97, 127
Long Sutton (Lincs.) 19, 62, 81, 100
Longford (Warwicks.) C25
Longsdon (Staffs.) 23
Longthorpe (Northants.) 64
Loughborough (Leics.) 209, 213
Loughton (Bucks.) 1, 3, 149, T8B
Lound: *see* Sutton cum Lound
Loversall (Yorks.) 13, 135
Lowdham (Notts.) 135
Lowe (Staffs.) 23
Lower Boddington: *see* Boddington, Upper & Lower
Lower Heyford (Oxon.) 7, 11
  see also Nether Heyford
Lower Lemington (Gloucs.) 92, 102, 125, 140
Lower Mitton (Worcs.) 14
Lower Radbourne (Warwicks.) C24
Lower Shuckburgh (Warwicks.) 36, 50, 93, 218, 300
Lower Slaughter (Gloucs.) 138
Lower Winchendon (Bucks.) 30
Lowesby (Leics.) 86, 91, 137
Lowick (Northants.), T14
  electricity 291
Loxley (Warwicks.) 22, 94
Lubbersthorpe (Leics.) 213
Lubenham (Leics.) 8, 37, 118, C4, C10, C11, C12, C13, C14
  electricity 289
  see also Thorpe Lubenham
Luddington (Northants.), electricity 291
Luddington (Warwicks.) 85, 94
Ludgershall (Bucks.) 262
Luffenham: *see* North Luffenham; South Luffenham
Luffield Abbey (Bucks.) 151, 189, 199, 234
Luton (Beds.) 23
Lutterworth (Leics.) 209, 213, T2
Lutton (Northants.), electricity 291
Lyndon (Rutland) 145, C15
Lynn, Wisbeach & Peterborough Midland Counties & Birmingham Junction Railway 45

Macclesfield (Cheshire) 23
McDonald, J.A. 207, 212, 215, 220, 230, 247, 252, 254, 263
Macneill, John 50, T8A, T8B
Maidford (Northants.), electricity 294
Maids Moreton (Bucks.) 51, C5, T4, T6
Maidwell (Northants.) 8, 42, 56, 66, C4
  electricity 294
Maltby (Yorks.) 13, 135
Manchester (Lancs.) 23
Manchester, Sheffield & Lincolnshire Railway 209, 213, 218, 221, 223, 226
Mann: *see* Coe & Mann
Mansergh, James 243
  see also Lawson & Mansergh
Manthorpe (Lincs.) 13, 31
Manton (Rutland) 15, 31, 43, 58, 132, 145, 164, C15
Marchington (Staffs.) 23
Marefield (Leics.) 86, 91, 135, 137
Marholm (Northants.) 12, 13, 15, 17, 52, 81, 97, 109, 208
Market Deeping (Lincs.) 16, 27, 49, 60, 89, 97, 127, 150, 156, 162, C14, C15
Market Deeping Railway 162
Market Harborough (Leics.) 8, 19, 28, 42, 67, C2, C3, C4, C11, C13, C14
  electricity 289
Market Harborough & East Norton Railway 88
Market Harborough & Melton Mowbray Railway 86, 91
Market Harborough, Melton Mowbray & Nottingham Railways 137

Market Overton (Rutland) 31, 40
Market Rasen (Lincs.) 31
Markham: see East Markham
Marlow: see Great Marlow
Marnham (Notts.) 13
Marsh Gibbon (Bucks.) 14
Marston (Lincs.) 13, 24, 31
 see also Long Marston; North Marston
Marston St Lawrence (Northants.) 14, 21, 38, 47, 53, 55, 92, 95
Marston Trussell (Northants.) C4
 electricity 296
Marsworth (Bucks.) 3, 14, C7
Martin, Johnson & Fox 13
Martin (Lincs.) 6, 13, 16, 49
Martinsthorpe (Rutland) 145, C15
Marton (Lincs.) 13
Marton (Warwicks.) 7
Maulden (Beds.) 23
Mawsley (Northants.) 220
 electricity 294
Maxey (Northants.) 13, 16, 19, 27, 49, 52, 57, 60, 89, 97, 111, 127, 150, 156, 162, 219, 268, C14, C15
Mears Ashby (Northants.)
 electricity 286
 gas 298
 Higham Ferrers water 237, 249
Medbourne (Leics.) 8, 28, 33, 37, 86, 88, 91, 135, 137, 142
Medmenham (Bucks.) 51
Melchbourne (Beds.), electricity 287
Melton Mowbray (Leics.) 15, 40, 41, 43, 58, 86, 91, 132, 135, 137
Mentmore (Bucks.) 3
Meppershall (Beds.) 20, 67
Meriden (Warwicks.) 1, T8B
Merton (Oxon.) 32
Metheringham (Lincs.) 6, 13, 16, 49, 97, 127, 160
Methley (Yorks.) 24
Metropolitan Railway 203
Mickleton (Gloucs.) 47
Mid-East England Electricity 295
Middle Claydon (Bucks.) 30, 38, 51, 53, 151, 189, 199, 234
Middle Littleton: see North & Middle Littleton
Middlethorpe (Yorks.) 13
Middleton, Reginald E. 249
Middleton (Northants.) 31, C13, C14
 electricity 296
 gas 304
Middleton Cheney (Northants.) 7, 11, 14, 21, 38, 47, 53, 55, 85, 92, 95, 102, 125, 140, 313
 rural district, electricity 294
Middleton Stoney (Oxon.) 14
Midland & Eastern Counties Railway 29, 39
Midland and Manchester, Sheffield & Lincolnshire Railways 135
Midland Counties & South Wales Railway 120
Midland Grand Junction Railway 51
Midland Railway, additional land at Isham 174
 additional land at Kettering and Rushton stations 123
 additional land at Peterborough 164
 additional land at Uffington Station 178
 additional land at Ufford 157
 additional land at Wellingborough 116
 alteration of line and branches near Wellingborough 63
 aqueduct at Peterborough 157
 branches from the Syston & Peterborough Railway 41
 Cransley Branch 157
 Cransley Branch extension 220
 extension from Leicester via Bedford to Hitchin 67
 extensions from Leicester to Hitchin 56
 extensions from to Northampton 56
 extensions to Huntingdon 56
 footpath at Higham Ferrers 212
 footpath at Irchester 212
 footpath at Isham Station 193
 Irchester and Raunds Branch 201
 junction with Stamford & Essendine Railway 65
 Kettering and Manton line 145
 land at Corby 284
 land at Cranford 247
 land at Desborough 284
 land at Helpston 215
 land at Kettering Station 230, 254
 land at Market Harborough 284
 land at Peterborough 207, 252, 254, 263
 land at Wellingborough 215
 land at Wellingborough Station 230
 Market Harborough Loop 157
 new footpath at Glendon and Barford 165
 new footpath at Irchester Station 182
 new footpath at Kettering 164
 new lines etc. at Hardingstone, Wellingborough, Peterborough, Finedon 83
 new road at Gretton 157
 Nottingham and Rushton lines 132
 Rushton and Bedford widening deviation 154
 Syston & Peterborough Railway 15, 58
 widening between Rushton and Bedford 136
 see also London & North Western Railway and Midland Railway
Midland Railway (Midland and Great Northern Railways Joint Committee), land at Dogsthorpe 276
Milcombe (Oxon.) 34, 92, 94a, 102, 138, 140
Milcote (Warwicks.) 29
Millbrook (Beds.) 107, 128
Miller, J. 301
 John 24, 29, 39, 59
Mills: see Lister & Mills
Milton (Northants.) 1, 8, 10, 64, 129, 134, 149, 152, 155, 216, 245, 246
 electricity 258
Milton (Oxon.) 94a, 138
 see also Great Milton
Milton Ernest (Beds.) 20, 23, 42, 43, 56, 67, 136, 154
Milverton (Warwicks.) 48, 50, C24
Mimms: see North Mimms; South Mimms
Missenden: see Great Missenden; Little Missenden
Misson (Notts.) 6, 24, 59, 89
Misterton (Leics.) 209, 213
Misterton (Notts.) 6, 13, 24, 59, 89
Mitton: see Lower Mitton; Upper Mitton
Mixbury (Oxon.) 18, 38, 196, 203, 209, 213
Moggerhanger (Beds.) 13
Molesworth (Hunts.) 33, 42, 56
Monk Sherborne (Hants.) 30
Monken Hadley (Middlesex) 13, 23, 24, 52
Monks Coppenhall (Cheshire) 23
Monks Kirby (Warwicks.) C21, C25, T2
Monk's Liberty (Lincs.) 13, 31, 89, 97
Monks Risborough (Bucks.) 51
Morborne (Hunts.) 13, 24
Morcott (Rutland) 8, 19, 37, 145
Moreby: see Stillingfleet with Moreby
Moreton Morrell (Warwicks.) 29
Moreton Pinkney (Northants.) 85, 94, 101, 105, 196, 203, 209, 213, 226, 233
Mortimer: see Stratfield Mortimer
Mortimer West End (Hants.) 30
Morton (Derbys.) 26
Morton (Lincs.) 6, 16, 49, 60, 89, 97, 127, C15
Mosley & Scrivener 248
Moss (Yorks.) 13, 97
Moulsoe (Bucks.) 107, 128, T5
Moulton (Northants.) 31
 electricity 258
 Northampton water 316
Moulton Park (Northants.), electricity 258
 Northampton water 281
Murrow (Cambs.) 79, 81
Muskham: see North Muskham; South Muskham
Myers-Beswick, W.B. 225

Nailstone (Leics.) 23
Napton on the Hill (Warwicks.) 11, 14, 29, 36, 50, 74, 93, 103, 190, 300
Naseby (Northants.) 33, C10
 electricity 294
Nassington (Northants.) 10, 64, 141
Naunton (Gloucs.) 138
Navenby (Lincs.) 13, 31, 89
Needingworth: see Holywell cum Needingworth
Neithrop (Oxon.) 14, 34, 47, 68, 85, 92, 95, 96, 98, 101, 102, 105, 125, 140
Nene, River 10a, 62, C16, C17, C19, C20, C22, C23
Nene Valley, drainage and navigation improvements 64, 72, 115
Nene Valley Waterworks 231
Nesham, William Joplin 74
Nether Broughton (Leics.) 40, 132, 135
Nether Heyford (Northants.) 3, 4, 29, 39, 48, 99, 103, 124, 133, 184, 185, 190, 245, 246, C7, T2, T8A, T8B
 electricity 290
 gas 292
 see also Lower Heyford
New Brentford (Middlesex) C77
New Sleaford (Lincs.) 89, 97, 127, 160
Newark upon Trent (Notts.) 13, 24
Newbold: see Owston & Newbold
Newbold Pacey (Warwicks.) 29
Newbold upon Avon (Warwicks.) 1, 3, C21, C25
Newborough (Northants.) 6, 19, 35, 54, 57, 59, 79, 81, 97, 100, 160, 210, 232
Newbottle (Northants.) 18, 21, 38, 47
Newent (Gloucs.) 125
Newington: see North Newington; South Newington
Newnham (Northants.) 29, 36, 50, 74, 78, 93, 112, 130, 161, 168, 176, 185, 188, 190, T8B, T13, T15
 electricity 290
Newport Pagnell (Bucks.) 61, 107, 128, C7, T4
 electricity 290
 gas 309
 rural district, electricity 290
Newport Pagnell Gas & Coke Co. Ltd 309
Newport Pagnell Railway 108, 119
Newstead (Lincs.) 75, 90
Newstead (Notts.) 26, 209, 213
Newton (Cambs.) 81
Newton (Lincs.) 24
Newton (Northants.) 31, 43, 145
 electricity 296
 gas 304
 see also Butley cum Newton; Cold Newton; Water Newton
Newton & Biggin (Warwicks.) 218
 Rugby water 285
Newton Blossomville (Bucks.) 25, 99, 103
 gas 309
Newton Bromswold (Northants.), electricity 287
 Great Ouse water 318
Newton Harcourt (Leics.) 8, 19, 20, 23, 67, C4

Newton Purcell (Oxon.) 18, 203, 209, 213
Newtown Linford (Leics.) 209
Nicholson, R. 32
Nixon & Dennis 86
Nocton (Lincs.) 6, 13, 16, 49, 97, 127, 160
Normanton le Heath (Leics.) 23
Normanton on the Wolds (Notts.) 132
Normanton on Trent (Notts.) 13
Normanton upon Soar (Notts.) 213
North & Middle Littleton (Worcs.) 14, 29
North & South Junction Railway 7
North Anston: see Anston, North & South
North Aston (Oxon.) 7
North Carlton (Lincs.) 89, 97, 160
North Claines (Worcs.) 85, 14, 22
North Clifton (Notts.) 24
North Elmsall (Yorks.) 13
North Hinksey (Berks.) 11, 14, 32
North Kilworth (Leics.) 37, C10, C12
North Leverton with Habblesthorpe (Notts.) 6, 24
North Luffenham (Rutland) 8, 15, C15
North Marston (Bucks.) 38, 51, 53
North Mimms (Herts.) 13, 24, 52
North Muskham (Notts.) 13
North Newington (Oxon.) 34, 47, 68
North Piddle (Worcs.) 85
North Scarle (Lincs.) 24
North Wingfield (Derbys.) 26
Northampton (Northants.) 8, 10, 25, 29, 31, 39, 42, 50, 56, 64, 66, 70, 77, 99, 103, 104, 106, 110, 124, 129, 133, 134, 149, 152, 155, 163, 167, 185, 186, 190, 235, C4, C7, C18, T5, T7
 borough extension and improvement 122
 cattle market 120a
 commons acquisition, new and widened streets 180
 electricity 200, 206, 258, 286, 290, 294
 gas 71, 121, 292, 298, 305, 306, 309
 street improvement and sewage outfall and irrigation works 122
 street widenings 274
 tramways 171, 179, 214, 242, 271, 274
 water 186, 281, 282, 283, 316
Northampton & Banbury & Metropolitan Railway 189
Northampton & Banbury Junction Railway 84, 102, 125, 129
Northampton & Banbury Railway 55, 95
Northampton & Blisworth Railway 152
Northampton & Daventry Junction Railway 133
Northampton & Daventry Railway 185
Northampton & Peterborough Railway 10
Northampton, Banbury & Cheltenham Railway 47
Northampton, Bedford & Cambridge Railway 25
Northampton Corporation, Commons, Streets, Roads Bill 180
Northampton Corporation Tramways 242, 271, 274
Northampton, Daventry & Leamington Railway 190
Northampton Electric Light & Power Co. Ltd 206, 258, 290, 293
Northampton Electric Power & Traction Co. Ltd 264
Northampton Gas Co. 298
Northampton Gas Light Co. 71, 121, 292
Northampton Gaslight Co. 305, 306, 309
Northampton Improvement Commissioners 122
Northampton, Lincoln & Hull Direct Railway 31
Northampton Tramways 214
Northamptonshire, petroleum filling stations 303
Northborough (Northants.) 16, 49, 89, 97, 127, 150, 156, 162
Northchurch (Herts.) 3, C7
Northern & Eastern Railway 6
Northill (Beds.) 13
Northorpe (Lincs.) 97
Norton (Herts.) 13, 24
Norton (Northants.) 1, 3, 4, 50, 78, 112, 130, 161, 168, 176, 185, 190, C7, C10, T2
 electricity 290
 see also Chipping Norton; East Norton; Over Norton
Norton Disney (Lincs.) 24
Norton juxta Kempsey (Worcs.) 14, 85
Norton Lindsey (Warwicks.) 36
Norwell (Notts.) 13
Norwood (Middlesex) 32, C7
Noseley (Leics.) 137, 142
Nostell: see Huntwick with Foulby & Nostell
Notgrove (Gloucs.) 138
Nottingham (Notts.) 26, 40, 43, 132, 209, 213

Oakham (Rutland) 15, 31, 41, 43, C15
Oakington (Cambs.) 6, 16, 49, 127
Oakley (Beds.) 20, 23, 42, 43, 56, 67, 136
Oakley (Bucks.) 32
Oakley (Northants.), gas 304
 see also Great Oakley; Little Oakley
Oddingley (Worcs.) 22
Oddington (Oxon.) 14
Odell (Beds.), electricity 287
 gas 309
Offchurch (Warwicks.) 50, 93, 103, 190, 300, C24
Offenham (Worcs.) 14
Offord Cluny (Hunts.) 13, 87
Offord D'Arcy (Hunts.) 13
Old Dalby (Leics.) 132
Old Hurst (Hunts.) 16, 49
Old (Northants.), electricity 290
 Northampton water 316
Old Sleaford (Lincs.) 49, 89, 97, 127
 see also Sleaford
Old Stratford (Northants.) C5, C7, C18
 gas 315
Old Stratford & Drayton (Warwicks.) 22, 29, 85, 94, 272
Old Union Canal (Grand Union) C10, C11, C12
Old Warden (Beds.) 13, 42, 67
Ollerton (Notts.) 135
Olney (Bucks.) 25, 61, 99, 103, 104, 106, 108, 110, 119, T7
 gas 309
 tramroads 205
Olney Gas Light Coke & Coal Co. Ltd 309
Olney Park Farm (Bucks.) 25
Ombersley (Worcs.) 14
Onley (Northants.) 14, C21, C25
Ordish & Le Feuvre 111
Ordsall (Notts.) 13
Orlingbury (Northants.) 31, T5
 electricity 288
 gas 307
Orton (Northants.), electricity 289
 gas 304
 Kettering water 243
Orton Longueville (Hunts.) 10, 24, 64
Orton Waterville (Hunts.) 10, 64
Osbournby (Lincs.) 97, 127
Oundle (Northants.) 10, 64
 electricity 291
Outwell (Cambs.) 100
Over (Cambs.) 6, 89, 97, 127, 160
Over Norton (Oxon.) 34, 94a, 98, 101, 102, 105, 138
Overbury (Worcs.) 47
Oversley (Warwicks.) 85, T5
Overstone (Northants.) 31
 electricity 286
Owen, W.G. 126
Owston (Yorks.) 13, 24, 89, 97, 135, 160
Owston & Newbold (Leics.) 86, 91, 135, 137
Owston Ferry (Lincs.) 13, 24, 97, 160
Owthorpe (Notts.) 40
Oxendon: see Great Oxendon
Oxenhall (Gloucs.) 125
Oxford (Oxon.) 7, 11, 14, 32, 53
Oxford & Rugby Railway 11
Oxford Canal C8
Oxford, Witney, Cheltenham & Gloucester Independent Extension Railway 32
Oxhey (Herts.) 1
Oxton (Notts.) 135

Packington, Great & Little 23, T8B
Padbury (Bucks.) 30, 38, 51, 151, 189, 199, 234
Paddington (Middlesex) 1
Papplewick (Notts.) 26
Parry, Edward 209, 213, 218, 221, 223, 226, 262
 Edward A. 233
Parson Drove (Cambs.) 46, 64, 79, 81
Passenham (Northants.) 30, 51, 192, C5, T4
 gas 310, 315, T8B
 see also Deanshanger; Old Stratford
Paston (Northants.) 6, 12, 13, 15, 16, 17, 49, 52, 79, 81, 89, 97, 100, 111, 127, 150, 156, 204, 276, C14, C15
Pattishall (Northants.), T8A, T8B
 electricity 290
 gas 305
Paulerspury (Northants.) 30, 51, 107, 128, T8B
 gas 298
Pauntley (Gloucs.) 125
Pavenham (Beds.) 20, 23, 42, 56, 67, 136
Paxton: see Little Paxton
Peakirk (Northants.) 13, 16, 19, 89, 97, 100, 111, 127, 210, 255, C14, C15
Pebworth (Worcs.) 14, 29
Peeir, Thomas C22
Perlethorpe cum Budby (Notts.) 135
Pershore (Worcs.) 14
Peterborough (Northants.) 6, 12, 13, 15, 16, 17, 35, 44, 45, 46, 49, 52, 54, 59, 62, 64, 72, 79, 81, 83, 89, 97, 109, 111, 127, 139, 146, 157, 158, 164, 183, 207, 219, 232, 236, 251, 252, 255, 263, 275, 279, C14, C15, C16, C17
 electricity 227
 gas 113, 143, 265
 Nene Valley Drainage and Navigation Improvements 115
 tramways 172
 water 109, 111, 150, 156
Peterborough & Nottingham Junction Railway 40
Peterborough Electric Light & Power Co. Ltd 227
Peterborough Gas Co. 265
Peterborough, Spalding & Boston Junction Railway 35
Peterborough Water Co. 150
Peterborough, Wisbeach & Sutton Railway 81
Peterborough, Wisbech & Lynn Junction Railway 46
Peterborough, Wisbech & Sutton Railway 100
Peterborough, Wisbech, Lynn & Boston Junction Railway 35
Pettifer, R. T11
Pick: see Everard, Son & Pick
Pickwell with Leesthorpe (Leics.) 137
Pidcock, John Hyde 122, 180

Piddington (Northants.) 99, 103, 169
  electricity 290
  gas 292
Piddington (Oxon.) 262
Pidley cum Fenton (Hunts.) 6, 89, 97, 127, 160
Pilham (Lincs.) 97, 160
Pillerton Hersey (Warwicks.) 14, 22, 85, 94
Pillerton Priors (Warwicks.) 14, 22
Pilsgate (Northants.) 13, 52, 76, 90, 143
Pilton (Northants.) 64
  electricity 291
Pilton (Rutland) 15, 145
Pinchbeck (Lincs.) 13, 19, 35, 49, 57, 160
Pinner (Middlesex) 1, 3
Pinvin (Worcs.) 14
Pinxton (Derbys.) 26
Pinxton Colliery Co. 209
Pirton (Herts.) 20, 107, 128
Pishill (Oxon.) 30
Pitchcott (Bucks.) 38, 51, 53
Pitsford (Northants.) 8, 42, 56, 64, 66, 70, 153
  electricity 290
  Northampton water 283, 316
Pitstone (Bucks.) 3, C7
Pixell, C. T9
Pleasley (Derbys.) 209
Ploughley (Oxon.), rural district 313
Plumpton (Northants.) 94
  electricity 294
Plumtree (Notts.) 132
Plungar (Leics.) 137, 142
Podington (Beds.) 20, 23, 29, 39, 43, 67
  electricity 287
  Higham Ferrers & Rushden water 249
Pointon (Lincs.) 6, 16, 49, 60, 97, 127, C15
Polebrook (Northants.) 64
  electricity 291
Pollington (Yorks.) 6, 13
Pontefract (Yorks.) 24
Ponton: *see* Great Ponton; Little Ponton
Port of Wisbech, Peterborough, Birmingham & Midland Counties Union Railway 44
Potsgrove (Beds.) T8B
Potter Hanworth (Lincs.) 6, 13, 16, 49, 97, 127, 160
Potterspury (Northants.) 30, 51, 107, 128, T8B
  gas 309
  rural district, electricity 290
Prescote (Oxon.) C24
Prestbury (Cheshire) 23
Preston (Rutland) 132, 145
Preston Bisset (Bucks.) 18, 196, 213
Preston Capes (Northants.) 22
  electricity 294
Preston Deanery (Northants.) 103, 169
  electricity 290
  gas 292
Preston on Stour (Gloucs.) 14, 85, 94
Prestwold (Leics.) 209, 213
Prichard, William B. 50
Princes Risborough (Bucks.) 51
Princethorpe (Warwicks.) 7
Priors Hardwick (Warwicks.) C24
Priors Marston (Warwicks.) 29
Provis, W.A. C15
Pulloxhill (Beds.) 23, 107, 128
Purdon, William 25
Pyrton (Oxon.) 30
Pytchley (Northants.) 20, 23, 31, 33, 42, 43, 56, 67, 80, 136, T5
  electricity 289
  gas 304

Quadring (Lincs.) 160
Quainton (Bucks.) 14, 18, 30, 38, 51, 53, 196, 203, 209, 213
Quarrendon (Bucks.) 14, 18, 38, 51, 196, 203
Quarrington (Lincs.) 49, 97, 127
Quatford (Salop) 262
Quatt Jarvis (Salop) 262
Queniborough (Leics.) 15
Quinton (Northants.) 169
  electricity 290
  gas 292

Radbourne: *see* Lower Radbourne; Upper Radbourne
Radclive (Bucks.) 38, 151, 189, 199, 234
Radford, William 64
Radford (Notts.) 26, 209, 213
Radford Semele (Warwicks.) 36, 50, 74, 93, 103, 190, 300, C6
Radstone (Northants.) 196, 203, 209, 213
Rampton (Cambs.) 89, 97, 127
Rampton (Notts.) 24
Ramsey (Hunts.) 6, 16, 49, 89, 97, 127, 160
Ranger, R. Apsley 66
Ranskill (Notts.) 13
Rastrick, J.U. 23, 31
Ratby (Leics.) 23
Raunds (Northants.) 10, 33, 42, 56, 64, 87, 201
  electricity 287
  gas 266
  tramroads 198, 239
Raunds Gas Light & Coke Co. Ltd 266
Ravensthorpe (Northants.) 186
  electricity 290
  gas 305
  Northampton water 281, 282
Ravenstone (Bucks.) 99, 103, 110
  gas 309
Ravenstone with Snibston (Leics.) 23
Reading (Berks.) 30, 51
Rearsby (Leics.) 15
Redbourn (Herts.) T8B
Redditch (Worcs.) 36, 85
Reepham (Lincs.) 31
Remenham (Berks.) 51
Remington, George 23
Rendel, James Meadows 49, 64
Rennie, George 40
  John 10a, 13, 23, 40
Retford: *see* West Retford
Richards, Edward 120, 125, 129, 140, 196, 209, 213
Rickmansworth (Herts.) 1, C7
Ridge (Herts.) 23, T8B
Ridgmont (Beds.) 107, 128
Ringstead (Northants.) 10, 33, 64, 87
  electricity 291
Rippingale (Lincs.) 6, 16, 49, 60, 89, 97, 127, C15
Risborough: *see* Monks Risborough; Princes Risborough
Riseley (Beds.) 29, 39
  electricity 287
Roade (Northants.) 1, 149, 167, 169, 216, 224, C7
  electricity 290
Robinson: *see* Shelford & Robinson 150
Rockingham (Northants.) 31, 132, 141, 145, C13, C14
  Corby water 299
  electricity 296
  gas 304
Rollestone (Staffs.) 23
Rollright: *see* Great Rollright; Little Rollright
Ropsley (Lincs.) 13
Ross, Alexander 221, 223, 226, 232, 251, 255, 268, 269, 275, 276
  see also Alexander & Ross
Rossington (Yorks.) 13, 59
Rotherby (Leics.) 15, 41
Rotherfield Peppard (Oxon.) 30

Rotherham (Yorks.) 13
Rothersthorpe (Northants.) 1, 10, 104, 129, 134, 152, 155, C7
  electric lighting 258
Rothley (Leics.) 213
Rothwell (Northants.) 20, 23, 33, 67, 136
  electricity 289
  gas 241, 304
  Kettering water 243
Rothwell Gas Light, Coal & Coke Co. Ltd 241
Rothwell Urban District Council 241
Rous Lench (Worcs.) 22, 85
Rousham (Oxon.) 11
Rowington (Warwicks.) C24
Rowlandson, C.A. 228, 233
Rowston (Lincs.) 6, 16, 49, 97, 127, 160
Roxholm (Lincs.) 89, 97, 127
Ruddington (Notts.) 209, 213
Rudyard (Staffs.) 23
Rufford (Notts.) 135
Rugby (Warwicks.) 1, 3, 11, 33, 37, 48, 149, 209, 213, 218, C21
  urban district council, general powers 285
  water 285
Rugby & Huntingdon Railway 33
Rugby & Stamford Railway 37
Rugby Urban District Council 285
Rumball, Thomas 45
Rushden (Northants.) 10, 64, 201
  electricity 259, 264, 277, 287 291
  gas 307
  Great Ouse water 318
  Higham Ferrers & Rushden water 249
  tramroads 198, 205, 239
Rushden & District Electric Supply Co. Ltd 287, 291
Rushden & Higham Ferrers District Gas Co. 307
Rushton (Northants.) 42, 56, 67, 123, 132, 135, 145
  electricity 289
  gas 304
Rushton Spencer (Staffs.) 23
Ruskington (Lincs.) 6, 16, 49, 97, 127, 160
Russel, Robert T14
Ryhall (Rutland) 13, 24, 52, 65, C15
Ryhill (Yorks.) 13
Ryther cum Ossendyke (Yorks.) 6
Ryton on Dunsmore (Warwicks.) 7

Saddington (Leics.) C4
St Albans (Herts.) 23, T8B
St Ives (Hunts.) 16, 49
St Marylebone (Middlesex) 3, 209, 213
St Neots (Hunts.) 13, 29, 39
St Pancras (Middlesex) 3, 4, 13, 24, 52
Saintbury (Gloucs.) 47
Salford (Beds.) 107, 128
Salford (Oxon.) 101, 102, 105
Salford Priors (Warwicks.) 22, 85, 94
Salperton (Gloucs.) 138
Sambourn (Warwicks.) 36, 85
Sampson, Brook 277
Sandal Magna (Yorks.) 13
Sandridge (Herts.) 23
Sandy (Beds.) 13, 24
Sarsden (Oxon.) 34
Saundby (Notts.) 6, 13, 24, 59, 89
Saunders, R.J.H. 176
Saunderton (Bucks.) 32, 51
Sawtry All Saints & St Andrew (Hunts.) 13, 24
Sawtry St Judith (Hunts.) 13, 24
Saxby (Leics.) 15, 40, 41, 58, 132
Saxelby (Leics.) 86, 91, 132
Saxilby cum Ingleby (Lincs.) 6, 13, 24, 160
Saxondale (Notts.) 137, 142
Scaftworth (Notts.) 13
Scaldwell (Northants.), electricity 290

gas 292
  Northampton water 316
Scampton (Lincs.) 89, 97, 160
Scarle: see North Scarle; South Scarle
Scopwick (Lincs.) 6, 16, 49, 97, 127, 160
Scothern (Lincs.) 31
Scott, H.E. 46, 61
Scotter (Lincs.) 97
Scredington (Lincs.) 89, 97, 127
Scrivener: see Mosley & Scrivener 248
Scrooby (Notts.) 13
Seaton, E.P. 231
Seaton (Rutland) 8, 19, 28, 31, 33, 37, 43, 141, 145
Seisdon: see Trysull & Seisdon
Selby (Yorks.) 13, 24
Sellon, Stephen 192, 198, 205, 239, 253
Sempringham (Lincs.) 6, 16, 49, 60, 89, 97, 127
Sevenhampton (Gloucs.) 138
Sharnbrook (Beds.) 20, 23, 29, 39, 42, 43, 56, 67, 136, 154
  electricity 287
Shawell (Leics.) 209, 213, 218
Shearman & Archer 267
Sheffield (Yorks.) 13
Sheffield, Nottingham & London Direct Railway 26
Shefford (Beds.) 20, 67
Shefford Hardwick (Beds.) 20, 42, 67
Sheldon (Warwicks.) 1, 3, T8B
Shelford, W. 156
Shelford & Robinson 150
Shelford (Notts.) 135, 137, 142
Shelswell (Oxon.) 18, 203, 209, 213
Shenley (Bucks.) T8B
Shenley (Herts.) 23, T8B
Sherborne (Warwicks.) 36
Sherfield upon Loddon (Hants.) 30
Sherington (Bucks.) 61, 119
Sherrard & Hall 13, 52
Sherriff, James C6
Shillington (Beds.) 13, 20, 107, 128
Shilton (Warwicks.) C25
Shinfield (Berks.) 30
Shiplake (Oxon.) 30, 51
Shipton (Gloucs.) 138
Shipton on Cherwell (Oxon.) 11
Shipton under Wychwood (Oxon.) 313
Shipway, J.H. 169
Shirburn (Oxon.) 30
Shirley, L.H. 169
Shottery (Warwicks.) 85
Shrewley (Warwicks.) 300, C24
Shuckburgh: see Lower Shuckburgh; Upper Shuckburgh
Shutford: see West Shutford
Shutlanger (Northants.) 30, 51, 169, C24
  electricity 290
  gas 309
Sibbertoft (Northants.), electricity 296
Sibford Ferris (Oxon.) 47, 68, 92, 125, 140
Sibford Gower (Oxon.) 47, 68, 92, 125, 140
Sibsey (Lincs.) 13
Sibson cum Stibbington (Hunts.) 10, 41, 64, 76, 90, 141
Silchester (Hants.) 30
Silk Willoughby (Lincs.) 89, 97, 127
Silsoe (Beds.) 107, 128
Silverstone (Northants.) 151, 189, 199, 234, T6
  (Northants.), electricity 294
Sinclair, Robert 79, 97
Skeffington (Leics.) 86, 91, 137, 142
Skegby (Notts.) 209
Skellingthorpe (Lincs.) 6, 13, 24, 89, 160
Skellow (Yorks.) 97
Skinnand (Lincs.) 13, 31
Skirbeck (Lincs.) 13, 49, 59
Slapton (Bucks.) C7

Slapton (Northants.) 21, 47, 55, 84, 120, C24
  electricity 294
Slaughter: see Lower Slaughter; Upper Slaughter
Slawston (Leics.) 28, 33, 86, 88, 91, 137, 142
Sleaford (Lincs.) 49, 97
  see also New Sleaford; Old Sleaford
Slipton (Northants.) 80
  electricity 291
Smeeton Westerby (Leics.) 8, 23, C4
Smith, George 114
  John C1
  Thomas R. 256
  Walrond 162
Snaith & Cowick (Yorks.) 6, 13, 24
Sneinton (Notts.) 132
Snelland (Lincs.) 31
Snibston: see Ravenstone with Snibston
Snitterfield (Warwicks.) 36
Solihull (Warwicks.) 1, 300, C21
Somerby (Leics.) 91, 137
Somerby (Lincs.) 13, 31, 97
Somersham (Hunts.) 6, 89, 97, 127, 160
Somerton (Oxon.) 7, 11, 14, 262
Sonning (Berks.) 30, 51
Soulbury (Bucks.) 3, C7
Souldern (Oxon.) 7, 11, 14, 32, 262, T10, T12
Souldrop (Beds.) 20, 23, 29, 39, 42, 43, 56, 67, 136, 154
  electricity 287
South Anston: see Anston, North & South
South Carlton (Lincs.) 13, 89, 97, 160
South Croxton (Leics.) 86, 91, 137
South Elmsall (Yorks.) 13
South Hinksey (Berks.) 11
South Kilworth (Leics.) 37, 285
  Rugby water 285
South Kirkby (Yorks.) 13
South Leverton (Notts.) 24
South Littleton (Worcs.) 14, 29
South Luffenham (Rutland) 8, 15, 19, 37, C15
South Midland, or Leicester, Northampton, Bedford & Huntingdon Railways 42
South Midland Counties Railway 8, 9
South Midlands & Southampton Junction Railway 30
South Mimms (Middlesex) 13, 23, 24, 52, T8B
South Muskham (Notts.) 13
South Newington (Oxon.) 34, 94a, 102
South Normanton (Derbys.) 26
South Scarle (Notts.) 24
South Stoke (Lincs.) 24, 31, 52
South Wales & Northamptonshire Junction Railway 68
South Wilford (Notts.) 40, 43, 209, 213
South Witham (Lincs.) 31
Southall (Middlesex) C7
Southam (Warwicks.) 7, 11, 14, 36, 50, 74, 93, 103, 190, C24, T1
Southill (Beds.) 20, 24, 42, 67
Southoe (Hunts.) 13, 24
Southorpe (Northants.) 13, 75, 90
Southwick (Northants.) 10, 64
Spalding (Lincs.) 13, 19, 35, 49, 57, 59, 160, C14
Spaldwick (Hunts.) 33, 42, 56, 87
Spanby (Lincs.) 89, 97, 127
Spernall (Warwicks.) 36
Spetchley (Worcs.) 85
Spratton (Northants.) 42, 56, 66, 186, C4
  electricity 290
  Northampton water 281
Springthorpe (Lincs.) 97
Stagsden (Beds.) 25, 103
Stainfield (Lincs.) 127

Stainforth (Yorks.) 6, 13, 97
Stainton by Langworth (Lincs.) 31
Stainton with Hellaby (Yorks.) 13, 135
Stamford (Lincs.) 8, 13, 15, 19, 28, 33, 40, 41, 52, 65, 75, 90, C13, C14, C15
  rural district, electricity 295, 297
Stamford Baron (Northants.) 15, 19, 40, 41, 52, 65, 69, 76, 90, C13, C14
  electricity 297
Stamford & Essendine Railway 65, 69, 75, 90
Stamford Junction Navigation C15
Stamford, Market Harborough & Rugby Railway 28
Stamford Rural, electricity 297
Stanford (Northants.) 33, 37, C10, C12
  Rugby water 285
Stanford upon Soar (Notts.) 209
Stanground (Cambs. & Hunts.) 6, 10a, 16, 35, 45, 46, 49, 62, 64, 72, 79, 89, 97, 115, 127, C14, C15
Stanion (Northants.) 145, T14
  electricity 296
  gas 304
Stanton: see Fraser & Stanton; Long Stanton All Saints; Long Stanton St Michael
Stanton & Newhall (Derbys.) 23
Stanton on the Woids (Notts.) 132
Stanton under Bardon (Leics.) 23
Stantonbury (Bucks.) 107, 128, C7, T4
Stanwick (Northants.) 10, 42, 56, 64, 87, 201
  electricity 287
  tramroads 198, 239, 260
Stapenhill (Staffs.) 23
Stapleford (Leics.) 15, 41, 43
Stapleford (Lincs.) 24
Starmore: see Westrill & Starmore
Stathern (Leics.) 137, 142
Staunton Harold (Derbys.) C1
Staunton (Gloucs.) 125
Staveley, Christopher C2, C3, C4
Staverton (Northants.) 29, 36, 50, T9, T11
  electricity 290
Steane (Northants.) 38, 53
Stears Brothers & Co. 113
Steeple Aston (Oxon.) 7, 11
Steeple Claydon (Bucks.) 30, 38, 151, 189, 196, 199, 203, 209, 213, 234
Stephenson, George 15
  George & Son 2, 3
  George Robert 66, 79
  Robert 4, 5, 10, 22, 36, 37, 41, 42, 48, 53, 56, 60, 61
  see also Hawkshaw, Fowler & Stephenson
Steppingley (Beds.) 107, 128
Stevenage (Herts.) 13, 24
Stevenson, Francis 141, 144, 147, 149, 153, 159, 163, 167, 170, 173, 175, 181, 184, 188, 197, 202, 211, 216, 217, 224, 229, 235, 245, 246
Stevington (Beds.) 99, 104
Stibbington: see Sibson cum Stibbington
Stillingfleet with Moreby (Yorks.) 6, 13, 24
Stilton (Hunts.) 13, 24
Stixwould (Lincs.) 13
Stockerston (Leics.) 137, 142
  Corby water 299
Stockton (Warwicks.) 7, 11, 36, 50, 74, 103, 190, 202, 300
Stoke (Warwicks.) C21
  see also South Stoke
Stoke Albany (Northants.) 135
  electricity 296
Stoke Bruerne (Northants.) 30, 51, 169, C7, C24
  electricity 290
  gas 309
Stoke Doyle (Northants.) 64

electricity 291
Stoke Dry (Leics. & Rutland) 132
  Corby water 299
Stoke Goldington (Bucks.) 103
Stoke Hammond (Bucks.) 1, 3, C7
Stoke Mandeville (Bucks.) 18
Stoke Poges (Bucks.) 262
Stoke Prior (Worcs.) 36
Stoke upon Trent (Staffs.) 23
Stondon: *see* Upper Stondon
Stone (Bucks.) 51
Stone (Worcs.) 36
Stoneleigh (Warwicks.) 1, 3, C21
Stonesby (Leics.) 132
Stoneton (Warwicks.) C24
Stonton Wyville (Leics.) 137, 142
Stony Stratford (Bucks.) 107, 128, 192, T8B
  gas 310, 315
Stony Stratford Gas & Coke Co. Ltd 310
Stopsley (Beds.) 23
Stotfold (Beds.) 13, 24
Stoulton (Worcs.) 14
Stourton (Warwicks.) 47, 68, 92, 102, 125, 140
Stow (Hunts.) 87
Stow (Lincs.) 97
Stow in Lindsey (Lincs.) 13
Stowe (Bucks.) 151, 189, 199, 234
Stowe (Lincs.) 13, 89
  see also Threekingham with Stowe
Stowe Nine Churches (Northants.) 3, 4, 29, 39, 48, 50, 99, 103, 124, 133, 185, 190, C7, T8A, T8B
  electricity 290
  ironworks 245
Stratfield Mortimer (Berks.) 30
Stratford le Bow (Middlesex) 127
Stratford-upon-Avon, Towcester & Midland Junction Railway 187
Stratford-upon-Avon & Midland Junction Railway 272
Streatley (Beds.) 23
Stretton Baskerville (Warwicks.) T2
Stretton on Dunsmore (Warwicks.) 7, T8B
Stretton on Fosse (Gloucs.) 47, 68, 92, 102, 125, 140
Stretton under Fosse (Warwicks.) C25
Strixton (Northants.) 29, 39, 61, 64, 108, 119
  electricity 288
  gas 312
  tramroads 198
Stubton (Lincs.) 13, 24
Stuchbury (Northants.) 21, 47, 55, 84, 120, C24
Studham (Beds.) 1
Studley (Warwicks.) 36, 85
Stukeley: *see* Great Stukeley; Little Stukeley
Sturton le Steeple (Notts.) 6, 24
Sturton (Lincs.) 13, 97
Styrrup (Notts.) 135
Sudborough (Northants.), T1
  electricity 291
Sudbrooke (Lincs.) 31
Sulby (Northants.) 33, C10
  electricity 294
Sulgrave (Northants.) 34, 95, 196, 203, 209, 213, 218, C24
Sulhampstead Banister (Berks.) 30
Surfleet (Lincs.) 13, 19, 49, 160
Sutterton (Lincs.) 13, 19, 49
Sutton (Cheshire) 23
Sutton (Northants.) 10, 13, 64, 75, 90, 109, C13
Sutton Basset (Northants.) 137, 142, C13, C14
  electricity 296
Sutton cum Duckmanton (Derbys.) 209
Sutton cum Lound (Notts.) 13
Sutton in Ashfield (Notts.) 209

Sutton on Trent (Notts.) 13
Sutton St Edmund (Lincs.) 19, 46, 79, 81
Sutton under Brailes (Warwicks.) 47, 68, 92, 125, 140
Swalcliffe (Oxon.) 47, 68, 92, 98, 101, 102, 105, 125, 140
Swansborough, W. C17, C19, C20, C23
Swaton (Lincs.) 6, 16, 49, 97
Swavesey (Cambs.) 16, 48, 49, 97
Swayfield (Lincs.) 13, 24, 52
Swerford (Oxon.) 34, 94a, 102
Swinderby (Lincs.) 24
Swineshead (Lincs.) 13, 49
Swinford (Leics.) 33, 37
  Rugby water 285
Swinstead (Lincs.) 13
Swithland (Leics.) 209, 213
Sykehouse (Yorks.) 6, 13, 97
Syresham (Northants.) 189, 199, 234
Sysonby (Leics.) 40, 86, 91, 132, 135, 137
Syston (Leics.) 15, 41
Syston (Lincs.) 13, 24, 31
Syston & Peterborough Railway 15, 54, 58
  see also under Midland Railway
Sywell (Northants.) 31, T5
  electricity 286
  gas 292
  Higham Ferrers water 237

Tackley (Oxon.) 7, 11
Tadmarton (Oxon.) 34, 47, 68, 92, 98, 101, 102, 105, 125, 140
Talke o'th'Hill (Staffs.) 23
Tallington (Lincs.) 13, 27, 49, 219, 278, C14, C15
Tansor (Northants.) 10, 64
Tanworth (Warwicks.) C21
Tardebigge (Worcs.) 36, 85
Tattershall (Lincs.) 13
Tattershall Thorpe (Lincs.) 13
Teeton (Northants.), electricity 290
  gas 305
  Northampton water 281
Teigh (Rutland) 15, 41, 58
Telford, Thomas C15, C21
Temple Bruer with Temple High Grange (Lincs.) 89, 127
Temple Grafton (Warwicks.) 22, 85, 94
Temple Hirst (Yorks.) 13
Tempsford (Beds.) 13, 24
Terrington St John (Norfolk) 45
Tettenhall (Staffs.) 262
Tetworth (Hunts.) 13
Teversall (Notts.) 209
Tewkesbury (Worcs.) 125
Thame (Oxon.) 30, 32
Theddingworth (Leics.) 37, C10, C12, C13, C14
Thenford (Northants.) 21, 38, 47
Thimbleby (Lincs.) 13
Thisleton (Rutland) 31
Thompson, J. Taylor 317
Thomson, James 50
Thornborough (Bucks.) 30, C5
Thornby (Northants.), electricity 294
Thorne (Yorks.) 13, 24
Thorney (Cambs.) 19, 44, 45, 46, 54, 64, 79, 160
Thorney (Notts.) 6, 24
Thornhaugh (Northants.) 41, 64, 75, 90, 109
Thornton, Francis Hugh 277
Thornton (Bucks.) C5
Thornton (Leics.) 23
Thornton le Fen (Lincs.) 13
Thorpe Achurch (Northants.) 10, 64, 153, 202
  electricity 291
Thorpe Arnold (Leics.) 15, 40, 41, 43, 58, 132, 137
Thorpe by Water (Rutland) 8, 19, 28, 31,

33, 37, 43, 145
Thorpe in Balne (Yorks.) 13, 97, 135, 160
Thorpe in the Fallows (Lincs.) 89, 97
Thorpe Langton (Leics.) 28, 33, 67, 137, 142
Thorpe Lubenham (Northants.) 37, 118
  electricity 296
Thorpe Malsor (Northants.) 20, 23, 33, 195
  electricity 289
  gas 304
  Kettering water 243
Thorpe Mandeville (Northants.) 34, 85, 95, 101, 105, 226
Thorpe Satchville (Leics.) 86, 91, 137
Thorpe Tilney (Lincs.) 49
Thrapston (Northants.) 10, 64, 80, 87
  electricity 291
  market 117
  rural district, electricity 259
Threekingham (Lincs.) 89, 97, 127
Thrup (Oxon.) 7, C7
Thrussington (Leics.) 15
Thurcaston (Leics.) 209, 213
Thurlaston (Warwicks.) T8B
Thurlby (Lincs.) 6, 16, 49, 60, 89, 97, 127, C15
Thurleigh (Beds.) 29, 39
Thurning (Northants.), electricity 291
Thursford (Norfolk) 276
Tibberton (Worcs.) 22, 85
Tibshelf (Derbys.) 26, 209
Tickencote (Rutland) 40
Tickhill (Yorks.) 13, 135
Ticknall (Derbys.) C1
Tiddington (Oxon.) 32
Tidmington (Warwicks.) 47, 68, 92, 102, 125, 140
Tiffield (Northants.) 21, 30, 34, 47, 51, 55, 84, 85
  electricity 290
  gas 292
Tilbrook (Hunts.) 87
Tilney All Saints (Norfolk) 45
Tilney St Lawrence (Norfolk) 45
Tilney with Islington (Norfolk) 45
Tilsworth (Beds.) T8B
Tilton (Leics.) 86, 91, 135, 137, 142
Timberland (Lincs.) 6, 13, 16, 49, 160
Tingewick (Bucks.) 38, 196
Tinsley (Yorks.) 13
Tinwell (Rutland) 8, 15, 19, 28, 33, C15
  electricity 297
Titchmarsh (Northants.) 10, 64
  electricity 291
Titherington: *see* Upton & Titherington
Tixover (Rutland) 28, 33, C13, C14
Todenham (Gloucs.) 47, 68, 92, 102, 125, 140
Todwick (Yorks.) 135
Toft (Cambs.) 29, 39
Tollerton (Notts.) 40, 43, 132
Tolmé, J.H. 108, 119
Tomlinson, Frank 281, 283
Torksey (Lincs.) 6, 13, 24, 160
Torworth (Notts.) 13
Tottenham (Middlesex) 13, 23, 24, 52
Towcester (Northants.) 21, 34, 47, 55, 84, 85, 94, 107, 128, 151, 169, 187, 189, 199, 234, 272, C24, T8B
  electricity 290
  gas 298, 305
Towcester & Buckingham Railway 199
Towcester & Hitchin Railway 128
Towcester Gas Co. 298
Towersey (Oxon.) 32
Treeton (Yorks.) 13, 135
Treswell (Notts.) 24
Tring (Herts.) 3, 14, 38, C7
Trumpington (Cambs.) 6, 29, 39, 127
Trysull & Seisdon (Staffs.) 262

Tugby (Leics.) 86, 91, 137, 142
Tur Langton (Leics.) 67
Turvey (Beds.) 25, 99, 103, 104
  gas 309
Turville (Bucks.) 30
Turweston (Bucks.) 38, 196, 203, 209, 213, 218
Tutbury (Staffs.) 23
Tutnall & Cobley (Warwicks.) 36
Twyford (Bucks.) 18, 196, 203, 209, 213
Twyford (Leics.) 86, 91, 135, 137
Twyford (Northants.) 64
Twywell (Northants.) 80
  electricity 291
Tydd St Giles (Cambs.) 62, 81
Tydd St Mary (Lincs.) 62, 81
Tyringham with Filgrave (Bucks.) 61, 119
  gas 309

Uffington (Lincs.) 13, 15, 19, 24, 49, 52, 65, 75, 90, 219, 116, C14, C15
Ufford (Northants.) 13, 15, 19, 27, 41, 52, 75, 90, 157
Ufton (Warwicks.) 36, 50, 74, 93, 103, 190, 300
Underwood, John 157, 164, 165, 173, 174, 178, 182, 193
Union Canal: see Leicestershire & Northamptonshire Union Canal
United Gas Co. 313
Upleadon (Gloucs.) 125
Upper Boddington (Northants.) 94
Upper Broughton (Notts.) 132
Upper Heyford (Northants.) 1, 50, 99, 103, 124, 133, 185, 190
  electricity 290
  gas 292
Upper Heyford (Northants.) T2
Upper Heyford (Oxon.) 7, 11, 14
Upper Hodnell (Warwicks.) 14
Upper Mitton (Worcs.) 14
Upper Radbourne (Warwicks.) 29, C24
Upper Shuckburgh (Warwicks.) 36, 93, 209, 213
Upper Slaughter (Gloucs.) 138
Upper Stondon (Beds.) 20
Upper Tadmarton (Oxon.) 47
Upper Tean (Staffs.) 23
Upper Winchendon (Bucks.) 30
Uppingham (Rutland) 132, 145
Upton & Titherington (Cheshire) 23
Upton Bishop (Herefs.) 125
Upton (Hunts.) 13, 24
Upton (Lincs.) 89, 97
Upton [near Northampton] (Northants.) 29, 39, 50, 64, 66, 75, 99, 103, 104, 124, 129, 133, 134, 152, 155, 185, 190
  electricity 258
Upton [near Peterborough] (Northants.) 13, 90
Upton Snodsbury (Worcs.) 85
Upton Warren (Worcs.) 36
  see also Headham cum Upton
Upwell (Cambs. & Norfolk) 100
Upwood (Hunts.) 16, 97
Urban Electric Supply Co. Ltd 297
Utting, Frederick J. 46
Uttoxeter (Cheshire) 23
Uxbridge (Middlesex) 32, 262, C7

Varley, John C2, C3, C4
Vignoles, Charles 19, C25

Waddesdon (Bucks.) 14, 18, 30, 38, 51, 53, 196, 203
Waddington (Lincs.) 13, 31, 89
Wadenhoe (Northants.) 64
  electricity 291
Wadworth (Yorks.) 135
Wakefield (Yorks.) 13

Wakerley (Northants.) 28, 33, 141, 147, C13, C14
Walcot [near Billinghay] (Lincs.) 16, 90
Walcot [near Falkingham] (Lincs.) 127
Walcot (Northants.) 90
Waldersea (Cambs.) 64
Wales (Yorks.) 135
Walesby (Lincs.) 31
Walgrave (Northants.) 31, T5
  electricity 290
  gas 292
  Northampton water 316
Walker, James C24
Walkeringham (Notts.) 6, 13, 24, 59, 89
Walpole St Andrew (Norfolk) 62, 81
Walpole St Peter (Norfolk) 45, 62, 81
Walsgrave on Sowe (Warwicks.) C21, C25
Walsworth (Herts.) 13
Walsoken (Norfolk) 45, 62, 81
Waltham on the Wolds (Leics.) 132
Walthamstow (Essex) 127
Walton (Northants.) 12, 13, 17, 79, 81, 97, 111, 127, 254, 275, 278
Walton (Warwicks.) 94
Walton (Yorks.) 13
  see also West Walton
Wanlip (Leics.) 213
Wansford (Northants.) 41, 64, 76, 90, 109
Wappenham (Northants.) 21, 34, 47, 55, 84, 95, 120, 196, 203, 209, 213, C24
  electricity 294
Warboys (Hunts.) 6, 16, 49, 89, 97, 127, 160
Warden: see Old Warden
Wardington (Oxon.) 14, 34, 85, 95, 101, 105, C24
Wardon (Worcs.) 85
Wargrave (Berks.) 51
Warkton (Northants.) 191
  Burton Latimer water 273
  electricity 289
  gas 304
Warkworth (Northants.) 7, 11, 14, 18, 21, 34, 38, 47, 53, 55, 68, 85, 92, 95, 96, 98, 101, 102, 105, 125, 140, 226, 313
Warmfield cum Heath (Yorks.) 13
Warmington (Northants.) 10, 64
Warmsworth (Yorks.) 13
Warndon (Worcs.) 22
Warrington (Bucks.) 61, 104, 108, 110, 119, 309, T7
  gas 309
Warwick 36, 48, 50, 300, C6, C24
Warwick & Braunston Canal C5
Warwickshire & London Railway 22
Washingborough (Lincs.) 6, 13, 16, 49, 97, 127, 160
Water Eaton (Bucks.) C7
Water Eaton (Oxon.) 7, 14
Water Newton (Hunts.) 13, 24, 64, 75
Water Stratford (Bucks.) 38
Watergall (Warwicks.) 14
Waterstock (Oxon.) 32
Watford (Herts.) 1, 3, C7, C18
Watford (Northants.) 1, 3, 4, 5, 149, 302, C10, T2
  electricity 290
Watlington (Oxon.) 30
Weedon Beck (Northants.) 3, 4, 22, 29, 36, 39, 48, 74, 78, 99, 112, 124, 133, 185, 188, C7, C21, T2, T8B
  electricity 290
  gas 292, 298, 306
Weedon Lois (Northants.) 21, 34, 55, 84, 95, 120
  electricity 294
Weedon & Daventry Railway 112
  see also under London & North Western Railway
Weedon & Leamington Railway 74

Weedon & Northampton Junction Railway 124
Weedon Gaslight Coke & Coal Co. 306
Weekley (Northants.) 31, 43, 131, 191
  Burton Latimer water 273
  electricity 289
  gas 304
Welbourn (Lincs.) 13, 31
Welby (Leics.) 86, 91, 132, 135, 137, T3
Welby (Lincs.) 13
Weldon (Northants.), gas 304
  see also Great Weldon; Little Weldon
Welford (Northants.) 33, 37, C10, C12
  electricity 294
  Rugby water 285
Welford on Avon (Warwicks.) 14, 22, 29
Welham (Leics.) 86, 88, 91, 137, 142
Welland, River, 65
Wellesbourne Hastings (Warwicks.) 29, 85, 94
Wellesbourne Mountford (Warwicks.) 29
Wellingborough (Northants.) 20, 23, 42, 43, 56, 61, 63, 64, 67, 83, 108, 116, 136, 215, 230
  electricity 240, 244, 280, 288
  gas 166, 250, 307, 312
  Great Ouse water 318
  Hind Inn T3
  rural district, electricity 264, 277
  tramroads 198, 205, 239, 260
  water 114
Wellingborough & District Tramroads 198, 205, 239, 253
Wellingborough Electric Supply Co. Ltd 280, 288
Wellingborough Gas Light Co. Ltd 166, 250, 307, 312
Wellingborough Urban District Council 244
Wellingore (Lincs.) 13
Wells-Owen & Elwes 190
Welton (Northants.) 1, 93, C7, C10, T2
  electricity 290
Welwyn (Herts.) 13, 24
Wendover (Bucks.) 18
Werrington (Northants.) 12, 13, 16, 17, 79, 81, 89, 97, 100, 111, 127, 150, 156, 208, 219, 236, C14, C15
West Bridgford (Notts.) 40, 43, 132
West Burton (Notts.) 6, 24
West Deeping (Lincs.) C14, C15
West Haddon (Northants.), electricity 290
  gas 305
West Langton (Leics.) 20, 67, C13, C14
West Retford (Notts.) 13
West Shutford (Oxon.) 92
West Walton (Norfolk) 45, 62, 81
West Wycombe (Bucks.) 32, 51
Westborough (Lincs.) 13
Westbury (Bucks.) 38, 151, 196, 203, 209, 213
Westcott (Bucks.) 30
Weston by Welland (Northants.) 8, 19, 28, 33, 37, 86, 88, 91, 135, 137, 142, 153, C13, C14
  electricity 296
Weston Favell (Northants.) 29, 31, 39, 64, 122, T5
  electricity 258
Weston on Avon (Warwicks.) 14, 22, 29
Weston on the Green (Oxon.) 14
Weston Subedge (Gloucs.) 47
Weston Turville (Bucks.) 38
Weston under Penyard (Herefs.) 125
Weston Underwood (Bucks.) 99, 103, 110, 309
  gas 309
Westrill & Starmore (Leics.), Rugby water 285
Westwick (Cambs.) 127
Wexham (Bucks.) 262

Whatborough (Leics.) 86, 91, 137, 142
Whatton (Notts.) 137, 142
Wheathampstead (Herts.) 23
Wheatley (Oxon.) 32
Wheatley (Yorks.) 13, 135
Whetstone (Leics.) 209, 213
Whichford (Warwicks.) 47, 68, 92, 101, 102, 105, 125, 140
Whilton (Northants.) 1, 3, 4, C7, T2
  electricity 290
  gas 305
Whissendine (Rutland) 15, 41, 43, 58
Whiston (Northants.) 10, 29, 39, 64
  electricity 290
  gas 309
Whiston (Yorks.) 13
Whitfield (Northants.) 151, 189
Whitnash (Warwicks.) 50, 103
Whittington (Gloucs.) 138
Whittlebury (Northants.) 51, 107, 128, 199, 234, T6, T8B
  electricity 290
Whittlesey (Cambs.) 6, 45, 62, 64, 72, 79, 81, 97, 115, 160
Whitworth, Robert C1
Wibtoft (Warwicks.) T2
Wicken (Northants.) 30, C5, T4
Wickenby (Lincs.) 31
Wickersley (Yorks.) 13
Widmerpool (Notts.) 132
Wiggenhall St Mary the Virgin (Norfolk) 45
Wigginton (Oxon.) 34, 92, 94a, 102, 138
Wigsthorpe: *see* Lilford cum Wigsthorpe
Wigston Magna (Leics.) 8, 19, 20, 23, 42, 56, 67, C4
Wilbarston (Northants.) 132, 135
  electricity 296
Wilby (Northants.) 64
  electricity 288
Wildsworth (Lincs.) 97
Wilford: *see* South Wilford
Wilkinson, James 210
Willen (Bucks.) 61, C7
Willenhall (Warwicks.) 3
Willersey (Gloucs.) 47
Willesden (Middlesex) 1, 3, 209, 213
Willey (Warwicks.) T2
Willingham (Cambs.) 6, 89, 97, 127, 160
Willingham (Lincs.) 13, 89, 97
Willington (Beds.) 13
Willmott, Russell 272
Willoughby (Warwicks.) 93, 209, 213, 218, C6, C12, C21, C25, T8B
  see also Silk Willoughby
Willoughby on the Wolds (Notts.) 132
Willoughton (Lincs.) 160
Wilshamstead (Beds.) 20, 23
Wilson, E. 138
  Edward 94a, 98, 151
  John 183
Wilsthorpe (Lincs.) 150, 156
Winchendon: *see* Lower Winchendon; Upper Winchendon

Wing (Rutland) 15, 31, 43, 132, 145
Wingerworth (Derbys.) 26
Winslow (Bucks.), rural district, electricity 294
Wintersett (Yorks.) 13
Winwick (Northants.) C10, C12
  electricity 294
Wisbech, River C19
Wisbech (Cambs.) 19, 44, 45, 46, 62, 64, 72, 79, 81, 100, 115, C17, C19
Wistow (Hunts.) 16, 49, 89, 97, 127, 160
Wistow (Leics.) 23, 42, 56, 67
Wistow (Yorks.) 6, 13, 24
Witham: *see* North Witham; South Witham
Withcote (Leics.) 86, 91, 137, 142
Withington (Lancs.) 23
Withybrook (Warwicks.) C25
Witney (Oxon.), rural district 313
Wittering (Northants.) 64, 76, 90, 109
Wiverton (Notts.) 137, 142
Wixford (Warwicks.) 22, 85
Wolfhampcote (Warwicks.) 36, 50, 74, 93, 103, 190, 202, 209, 213, 216, 300, C25, T1
Wolford: *see* Great Wolford; Little Wolford
Wollaston (Northants.) 10, 29, 39, 61, 64, 108
  electricity 288
  gas 267, 312
  Higham Ferrers & Rushden water 249
  tramroads 198
Wollaston (Worcs.) 14
Wollaston Gas Coal & Coke Co. Ltd 267, 312
Wolstanton (Staffs.) 23
Wolston (Warwicks.) 1, 3, 4, 7
Wolvercote (Oxon.) 7, 11
Wolverhampton (Staffs.) 262
Wolverley (Worcs.) 14
Wolverton (Bucks.) 1, 3, 4, 107, 128, 149, C7, T4, T8B
  electricity 290
  gas 310
Wolverton (Warwicks.) 36
Wolverton & Stony Stratford Tramways, Deanshanger Extension 192
Wolverton & Stony Stratford Tramways Co. Ltd 192
Wolvey (Warwicks.) T2
Wombourn (Staffs.) 262
Wooburn (Bucks.) 32
Wood Walton (Hunts.) 13
Woodall, Corbet & Son 265
Woodend (Northants.) 94
  electricity 294
Woodford (Cheshire) 23
Woodford cum Membris (Northants.) 22, 85, 94, 209, 213, 218, 223, 228, 233, 301
  electricity 294
  gas 313
Woodford near Thrapston (Northants.) 10, 33, 64, 80, 87, 175
  electricity 291
  gas 307
Woodhall (Lincs.) 13
Woodham (Bucks.) 14
Woodhouse, William 68
Woodhurst (Hunts.) 16, 49
Woodnewton (Northants.) 64
Woodstone (Hunts.) 10, 13, 24, 44, 62, 64
Woolscott (Warwicks.) C6, T8B
Woolstone: *see* Great Woolstone; Little Woolstone
Wootton (Northants.) 1, 8, 48, 64, 103, 129, 134, 149, 152, 155, 216, C7
  electricity 200, 258
Wootton Bassett: *see* Cricklade & Wootton Bassett
Wootton Wawen (Warwicks.) 36
Worcester (Worcs.) 14, 85
Worcester & Broom Railway 196
Worksop (Notts.) 135
Wormington (Gloucs.) 47
Wormleighton (Warwicks.) 7, 14, 22, 85, 94, C24
Worthington, W.B. 276
Wothorpe (Northants.) 40, C13, C14, C15
  electricity 297
Wotton Underwood (Bucks.) 30, 262
Woughton on the Green (Bucks.) 1, 3, 61, 149, C7
Wragby (Yorks.) 13
Wright, John 91
Wroot (Lincs.) 6, 13, 24, 97, 160
Wrottesley (Staffs.) 262
Wroxall (Warwicks.) C24
Wyberton (Lincs.) 13, 19, 49
Wycombe: *see* Chepping Wycombe; High Wycombe; West Wycombe
Wyfordby: *see* Brentingby & Wyfordby
Wyken (Warwicks.) C21
Wymington (Beds.) 20, 23, 42, 43, 56, 67, 136, 154
  electricity 287
  Higham Ferrers & Rushden water 249
Wymondham (Leics.) 15, 40, 41, 58
Wymondley: *see* Great Wymondley; Little Wymondley
Wyton (Hunts.) 16

Yardley Gobion (Northants.) 51, C7
  gas 309
Yardley Hastings (Northants.) 25, 99, 104, 110, T7
  electricity 290
  gas 309
Yardley (Worcs.) 1, 3, T8B
Yarnton (Oxon.) 11
Yarwell (Northants.) 41, 64
Yaxley (Hunts.) 13, 24
Yelden (Beds.), electricity 287
Yelvertoft (Northants.) C10, C12
York (Yorks.) 6, 13, 24